THE SCIENCE OF

WOMEN IN HORROR

THE SCIENCE OF

WOMEN IN HORROR

THE SPECIAL EFFECTS, STUNTS, AND TRUE STORIES BEHIND YOUR FAVORITE FRIGHT FILMS

Meg Hafdahl &
Kelly Florence

AUTHORS OF
The Science of Monsters

Skyhorse Publishing

Skyhorse Publishing books may be purchased in bulk at special discounts for sales promotion, corporate gifts, fund-raising, or educational purposes. Special editions can also be created to specifications. For details, contact the Special Sales Department, Skyhorse Publishing, 307 West 36th Street, 11th Floor, New York, NY 10018 or info@skyhorsepublishing.com.

Skyhorse® and Skyhorse Publishing® are registered trademarks of Skyhorse Publishing, Inc.®, a Delaware corporation.

Visit our website at www.skyhorsepublishing.com.

10 9 8 7 6 5 4 3 2

Library of Congress Cataloging-in-Publication Data is available on file.

Cover design by Peter Donahue and Daniel Brount
Cover photograph by gettyimages

Print ISBN: 978-1-5107-5174-3
Ebook ISBN: 978-1-5107-5176-7

Printed in China

We dedicate this book to the women who slayed monsters long ago.

CONTENTS

INTRODUCTION

It is said that a nineteen-year-old girl invented the modern horror genre. *Frankenstein*, borne of the imagination of Mary Shelley, posed a vital question to its readers, one that exists in nearly all horror movies today: Who is the true monster? The creator? The creature? Often, this search leads us down dark and terrifying corridors illuminated only by knowledge and science. These are the investigations that thrill us, that further our love of horror.

How do we, as women, reconcile the sometimes violent, misogynistic nature of the horror genre? How can we be fans of the problematic ways in which women are portrayed? The answer is in seeking out those films and television shows that give us the flawed, complicated, and *real* women we want to see on screen.

What do we want from our female characters? They don't need to be perfect. They don't necessarily need to kick ass. They don't need to make all the right decisions. We want to see a reflection of ourselves. *The Science of Women in Horror* explores the way women have been trailblazers and creators within the genre from its infancy and digs into the archetypes, social science, and history behind horror itself, all while speaking with such notable female horror legends as Dee Wallace (*Cujo, E.T.*), Deborah Voorhees (*Friday the 13th: The New Beginning*), and Alice Lowe (*Black Mirror: Bandersnatch, Sherlock*).

Join us as we discover the scientific proof that ghouls rule!

SECTION ONE
THE MOTHER

CHAPTER ONE

PREVENGE

Year of Release: 2016	
Director: Alice Lowe	
Writer: Alice Lowe	
Starring: Alice Lowe, Kate Dickie	
Budget: $104,000	
Box Office: $94,100	

"It's alive!" Frankenstein's monster was birthed out of science and curiosity, fascination and discovery. Mary Shelley's "hideous progeny" of a novel was both the birth of the horror genre and of her, as a writer. Women are mothers in various forms. Whether they give literal birth or bring forth an artistic project, women are creators.

Pregnancy, and being a parent, are things that women have struggled with since the dawn of time. Women anguish over getting pregnant. If they can't, they feel that they have failed. If they can, they fear that they are doing things wrong and compare themselves to others. This phenomenon may be more prevalent in the time of social media but has existed since the Victorian era. According to the authors of *You're Doing it Wrong! Mothering, Media, and Medical Expertise*:

> The failed femininity thing is a constant message. You are always failing your femininity in a different way, so it's a moving target. Some things have changed in that, in the Victorian age, the idea that you would work outside the home and you would not be a stay-at-home mom, is failed femininity. Now the conversation is much more barbed, because there are many women and people that identify as women that [work outside the home]. There can be defensiveness around this, and then the conversation looks

like it's happening between stay-at-home moms and moms who work.[1]

These complicated feelings of fear, regret, and worry regarding motherhood are present in horror literature, film, and television. Exploring the fears about pregnancy, specifically, are brought up in the first film discussed in this book.

"There is something terrifying and godlike about making life, and it's about time a woman owned that on screen."[2] Anyone who has been pregnant or even been near a pregnant woman can tell you how strange it is to see the baby move, visibly, for the first time. Although it could be considered the most natural occurrence in nature, pregnancy seems almost otherworldly. A parasitic, unknown being inhabits a woman's body and feeds off of her, influences her moods and emotions, and physically affects her entire life. Dramatic description? Perhaps. But for the mother in the movie *Prevenge* (2016), the baby affects all of these things and more.

Alice Lowe, who wrote, directed, and starred in *Prevenge* as the lead character, Ruth, completed filming in only eleven days of shooting. Any one of those accomplishments can be considered an incredible feat, but it's even more extraordinary when we learn that she was actually pregnant while filming. The story follows a pregnant woman who is getting revenge on those she believes killed the father of her baby. This plot alone would be compelling enough, but add in the element of the unborn baby talking to and directing the mother to kill, and it makes for a horror movie like no other.

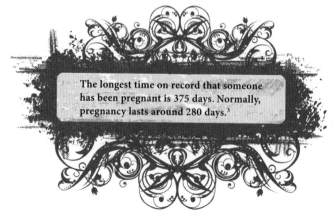

The longest time on record that someone has been pregnant is 375 days. Normally, pregnancy lasts around 280 days.[3]

There are numerous examples of women filming while pregnant. Madonna famously filmed the musical *Evita* (1996) while pregnant, as did Gillian Anderson while working on *The X-Files* in 1994. Women on TV shows who are filming while pregnant in real life are often seen hiding their stomachs with laundry baskets or under bulky clothing. Neither of these tricks was needed for *Prevenge*. The character of Ruth was written to be very pregnant throughout the film. Lowe purposely shot quickly to avoid any continuity issues with her belly size and to avoid getting too tired while on set.

What inspired her to make this movie? Lowe was continuing to audition for projects and had aspirations to direct a feature-length film, but opportunities just weren't coming to her. "If you're a woman over thirty-five, no one is going to hand you a free pass."[4] Things fell into place, and with funding from a production company guaranteed, Lowe realized she could use her pregnancy to her advantage and base the movie around it. She pitched it as a female *Taxi Driver* (1976):

> Female characters are always mothers or girlfriends who provide some sort of network for the hero, who then goes out and does whatever he wants. But what about a woman who's cut off from society? Ruth's philosophy is that society is selfish, and collectively [her victims] made a bad decision to destroy the love of her life [her husband, whose death we learn about via creepy flashback] and ruined the future of her baby.[5]

The script was completed in two weeks. While writing the script, Lowe wondered if her character would be likable enough but noted that the same is not often asked of male characters. A woman's focus in the media and in many other industries is often on her likability instead of her strengths, whereas the likability of a man is seldom brought up. No one asked, "Is Travis Bickle [from *Taxi Driver*] likable enough? Will he reflect badly on men?"[6]

We had the opportunity to interview Alice Lowe about the making of the film and about her experience being a mother.

Kelly: **"During the time of filming, did you find that people in life, or on set, treated you differently when you were pregnant?"**

Alice Lowe: "Occasionally people would bring me a chair to sit on and I'd be like 'why?' Then I'd remember I was pregnant. I frequently forgot because I was so engaged with the filming and was enjoying myself so much. Then toward the end I was getting really huge and really tired and I did think, *Oh yes, I would quite like to just lie down and sleep.*"

Kelly: **"I felt that way when I was pregnant, and I wasn't doing nearly as much as you were!"**

Alice Lowe: "When we finished filming, I went to some parenting classes, and the teacher asked everyone, 'Have you been putting your feet up?' I was like, 'No, I've been filming up a cliff in Wales.' I actually think it helped distract me from that last bit of pregnancy, which can be really boring. You're just sitting around thinking about your aches and pains. I really had to force myself to adjust to the idea of motherhood, though, as I'd just been being really active and running around. (Hence the parenting classes.) I said to my partner, 'I know nothing about babies.' That was about a month before I gave birth! People on set were brilliant, though. There was neither huge paranoia and fear, which may have set me on edge, nor any pushiness whatsoever. I didn't do anything I felt uncomfortable doing. And I could not have done it without the support of an understanding and trusting team. Goodness, how nervous it must have made everyone else watching a pregnant director—they certainly didn't show it!"

Meg: **"Were there any expectations about motherhood, in your own life, that didn't turn out the way that you thought?"**

Alice Lowe: "I certainly thought it was going to be a lot worse. Which is why the film is so dark! I really felt I was going to lose my identity. Which is why the film centers on grief and loss. The idea that the 'old you' has died. And you have no idea who or what is going to take over. There certainly is a shift in how you see yourself and your priorities, but post birth/pregnancy, I still feel like the same person, albeit one who has had some different experiences. What I didn't expect was how nice having a baby is. I think I went in thinking very cynically that I am quite a headstrong, independent person. I was fearful I would resent the baby. But actually, I was really surprised how much I loved motherhood. I think there are so many narratives supporting the idea that your life is over when you have kids (for women anyway). And actually, I was very pleasantly surprised. But I do think this is because I had shaped my own narrative by keeping working throughout, which many people would frown upon. I took my baby everywhere with me, on a world tour with the film, and it was a very fulfilling experience. I thought, *Whoever told me I had to give anything up, they were just plain wrong!* Now I realize though that filmmaking with a toddler is probably much harder, as opposed to their being in your tummy. It gets harder as they get older and have more of their own lives and commitments. The baby bit is actually the easy bit in terms of taking them with you. And nobody really tells you that."

Kelly: **"Dark humor can be tricky, but it's perfectly executed in *Prevenge*. How did you find that balance when writing the script?"**

Alice Lowe: "It's kind of my thing, so I guess it's something that either comes naturally to me or that I've honed over years of working as a writer/actor. But to me, you can sell anything as long as there's truth. I mean by this that it doesn't matter if you're playing comedy, or a goblin, or a dragon owner, or a taxi driver, as long as you bring some human truth to it, people can identify with it. So, I always play everything straight and with really high stakes, regardless of whether it's comedy or not. People don't know they are funny; everyone is living their own tiny struggle and tragedy. So, I just played as much truth in the performance as I could and also cast it with that in mind. When you know what sort of performers and performance you're going for, it gives you faith in the script. The script can be relatively outlandish as long as people are going to come in and play it with truth. My thing with *Prevenge* was, whether you liked it or not, you were going to care for the character. Even if this was through contrition and the audience was going to be gradually worn down. That despite what she's doing, she's a vulnerable, flawed human being. If you can nail that, then you've sold it to the audience. Not everyone buys, though, so it can be risky. I like to combine polar opposites though, a warm actor doing cold things, for example. You pull people this way and that. You make the audience think they hate someone, then add a detail that makes them love them. Before they know it, they're hooked in. It probably comes from my theater background. The audience is king/queen. Also, I would say that people often forget that all films are *fantasy*. Anything you see in a film is symbolic or representative of the mind: fears, hopes, dreams. So, *Prevenge* is an idea, not a documentary. It's a question, not an answer. As all films are."

Meg: "**We loved your performance in *Black Mirror: Bandersnatch*. Can you tell us about that experience?**"

Alice Lowe: "I'm a big fan of *Black Mirror,* so I just held my breath when I found out there was a chance that I was going to be in it. Then, when I found out it was a 'choose-your-adventure' format, it was so exciting, but really very mind-boggling. And you're not sure how it will work or *if* it will work. But then, you're working with geniuses, so you put your faith in them and just feel fortunate to be working in a new experimental way. I think I was very lucky with that particular character because my scenes were all in one place, in one situation, in one chair usually! It made life easier for me. The lead actor was acting in several different situations, different narrative threads, different outcomes. Fionn [Whitehead] seemed to have boundless energy, but I think it could have been a real mindf*ck! The team, director, and producer were all over the different threads, so I just trusted in them and relaxed and had a brilliant time. I particularly enjoyed doing the fight sequence!"

Meg: "**That was my favorite part!**"

Kelly: **"What are your current or future projects? Are you planning to write and/or direct again?"**

Alice Lowe: "I'm due to film my second feature, *Timestalker*. It's a reincarnation rom-com. I'm also working on a Delia Derbyshire biopic. She was an electronic music pioneer who cowrote the *Doctor Who* (1963) theme tune. I'm really lucky that I'm getting lots of opportunities to develop lots of different types of projects. I'm also developing television ideas. It seems that in the UK it's finally hitting home the wealth of talent, especially female talent, available. Once upon a time, I would be pitching ideas, e.g., *Sightseers* (2012), to television and would be told 'too violent,' or 'we can't mix comedy and horror,' etc. But since *Killing Eve* (2018–present), suddenly people are looking for exactly what it is I've been pitching for the last fifteen years. It feels like everything is suddenly very open. A lot of this is due to the competition of Netflix, for example. And the hunger for the new, diverse, exciting, shocking, and boundary-pushing. I don't know if it can last, but it certainly feels like a sea-change."

Lowe described the process of writing, directing, and going through postproduction on the film as similar to pregnancy and birth. While female writers and directors in the film industry are becoming more common, they still only made up 8 percent of directors in Hollywood and 16 percent of writers in 2018.[7]

Giving birth to evil is a concept that has been explored in several horror movies including *Rosemary's Baby* (1968), *Devil's Due* (2014), and *The Prodigy* (2019) and gives viewers an interesting perspective of the mother as a protector but also perhaps as a killer. Pregnant women are, unfortunately, often the victims of murder, and it is the leading cause of death during pregnancy. Statistics show that 20 percent of women who die while pregnant are the victims of a violent crime.[8] Having a pregnant woman be the killer in a horror movie subverts our expectations. Any thought that the character of Ruth is weak or docile quickly evaporates as we see her exact her revenge on the perceived villains throughout the film. The blend of horror and dark comedy makes for a shocking viewing.

"You have absolutely no control over your mind or your body anymore. This one does. . .," the midwife says to Ruth as she pats her pregnant belly during a scene in *Prevenge*. How does pregnancy influence emotions and mental health? The character of Ruth in *Prevenge* could be diagnosed as suffering from prepartum anxiety and grief. How does this affect real pregnant women? Grief can affect pregnancy through its impact on hormone balance and production, cause an imbalance in serotonin production, and raise cortisol levels.[9] Serotonin is often seen as a contributor to feelings of well-being and happiness but also affects learning and memory processes in the brain. Cortisol is a steroid hormone and acts as our natural stress or alarm system. Having these hormones out of balance can not only affect the mother during pregnancy, but the baby, as well. Grief during pregnancy could also cause physical symptoms in the mother including aches, pains, sleep problems, and digestive issues. Experts recommend seeing a doctor and therapist if pregnant and experiencing grief.

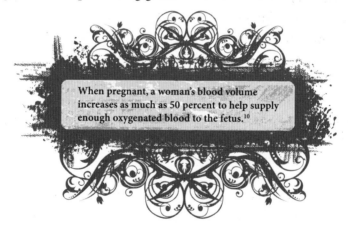

When pregnant, a woman's blood volume increases as much as 50 percent to help supply enough oxygenated blood to the fetus.[10]

There are a number of mental health issues that could arise after a woman gives birth. Almost half of all cases of mental illness among women, before and after they give birth, go undetected. Many of those who are diagnosed do not get the treatment they need. A condition affecting one in nine women in the United States is postpartum depression. Symptoms can range from mood swings and crying spells to extreme depression and suicidal thoughts. In 2019, the FDA approved the first

drug for mothers suffering from severe postpartum depression. It can be administered through an IV over sixty hours and requires a hospital stay. The drug, however, can cost between $20,000 and $30,000 and is not necessarily approved by insurance companies.[11] Postpartum depression can develop into postpartum psychosis, which has more severe symptoms including obsessive thoughts, hallucinations, delusions, and paranoia. This is much rarer, occurring in only one to two births out of every thousand. The most significant risk factors for developing postpartum psychosis are having a personal or family history of bipolar disorder and having had a previous psychotic episode.[12]

Cesarean sections are performed in about one in three births in the United States.

Birth itself has been portrayed in horror movies before, and *Prevenge* is no exception. A cesarean section delivery is shown in gory detail with bright red blood filling the frame. Cesarean sections are performed in about one in three births in the United States. The umbilical cord becomes a symbol of being tethered to another, just like Ruth's husband during the climbing accident in the movie. Making a cut, symbolized by the C-section, is oftentimes a choice between life and death, and Ruth realizes her husband's death may have been necessary to save others' lives. Other birth scenes in horror movies occur in *Alien* (1979), *The Fly* (1986), and *A Quiet Place* (2018). Each has their own horror attached to a moment that may be horrific enough on its own.

Whether it's a healthy baby being born in a postapocalyptic world like in *Dawn of the Dead* (1978), or telekinetic babies being brought into a peaceful town like in *Village of the Damned* (1960), one thing is certain: motherhood and pregnancy are rife topics to explore in any genre, but especially horror.

CHAPTER TWO
THE BABADOOK

Year of Release: 2014	
Director: Jennifer Kent	
Writer: Jennifer Kent	
Starring: Essie Davis, Noah Wiseman	
Budget: $2 million	
Box Office: $7.5 million	

Marie Curie was the first woman to win a Nobel Prize after conducting groundbreaking research on radioactivity and discovering radium and polonium. A film, *Madame Curie* (1943), explored her life and ends with the true incident of her husband dying tragically in a road accident. Marie Curie was a mother who experienced catastrophe much like the mother experiencing grief in *The Babadook* (2017).

Amelia (Essie Davis) becomes a widow when her husband is killed in a traffic accident on the way to the hospital while she is in labor. The trauma and grief that she experiences manifest themselves as a creature that affects her and her son, Sam (Noah Wiseman), throughout the film. Jennifer Kent, who wrote and directed the film, said, "I'm not a parent, but I'm surrounded by friends and family who are, and I see it from the outside . . . how parenting seems hard and never-ending. I think [the movie has] given a lot of women a sense of reassurance to see a real human being up there. We don't get to see characters like her that often."[1]

Feeling detached or not bonded to a child is a real issue that many mothers face. Beyond the feelings of postpartum depression, women report feeling indifferent, ambivalent, or even dislike toward their own children but often don't share these feelings due to fears of being the outliers of such thoughts. True neglect can also lead to a plethora of problems. Studies find that children who have spent their early childhoods in institutional settings in which they received care, but not love, develop weakened immune systems and have problems with social interaction.[12]

The topic of mothers killing their children is explored in *The Babadook*, and Kent acknowledged it while promoting the film. "Now, I'm not saying we all want to go and kill our kids, but a lot of women struggle. And it is a very taboo subject, to say that motherhood is anything but a perfect experience for women."[3] This theme is explored in horror movies such as *Oculus* (2013) and *Lights Out* (2016). There's nothing more terrifying than having someone you love and trust turn on you. The one person most children feel safest with is their mother. When she becomes the villain, whether by otherworldly means or by her own volition, the horror is real. This concept isn't new, of course. The Greek myth of Medea explores a mother killing her own children in order to seek vengeance on her former husband.

The Babadook was inspired by the real-life childhood fears of one of Kent's friends. The woman's son "was traumatized by this monster figure that he thought he saw everywhere in the house. So, I thought, 'What if this thing was real, on some level?'"[5] The monster as a physical manifestation of a parent's repressed grief makes *The Babadook* unique, emotional, and terrifying. The movie

The mother and son in *The Babadook* may be suffering from folie à deux, or a madness shared by two, in the film.

also explores what could be the concept of folie à deux, or a madness shared by two. This condition has been documented throughout history and has caused numerous people to commit murder or die by suicide. In *The Babadook*, Amelia doesn't immediately believe what her son is telling her about the creature, a common recurring theme in horror movies. But as she begins to see and believe the same things as her son, it becomes a shared experience.

In the 19th century, the first reported case of folie à deux involved a married couple who shared delusions that people were entering their house, spreading around dust and fluff, and wearing down the couple's shoes.[6]

Women in the horror genre are often portrayed as not feeling in control of their own minds or as having a fear of going mad. Kent said:

> I didn't want to portray Amelia as this crazy woman from the get-go . . . Often, women who are crazy are demonized in films, because we look at them from the outside. I really wanted to experience what it was like to go down that slippery slope from the inside. I wanted to create a woman who was really just struggling, while also pointing out that this monster [exists] within everyone.[7]

Mothers are often seen as a stable constant in their family's lives, but societal pressure and expectations, coupled with a heightened or intense situation, push them to the breaking point. It's this point that is explored in horror films such as *The Babadook*.

Another taboo subject that's been brought up in relation to the film is the idea of not loving motherhood. As Heather Havrilesky said in an opinion piece in the *New York Times*:

> When I hear someone telling an expectant mother that having a baby will turn her into a new person, I can't help but imagine a pathologically optimistic weather forecaster brightly warning

that an oncoming tornado is about to give a town "an extreme makeover." Becoming a mother doesn't change you so much as violently refurbish you, even though you're still the same underneath it all.[8]

Many women don't experience the immediate, euphoric feeling of bonding with their children, and some never do. Being a parent is a difficult and often thankless job that can be isolating, especially when one feels that they are alone or wrong in their feelings about it. Having open and honest communication with loved ones, and having representation of all parenting styles and experiences in media, can only better inform the public about these differences and help people realize they are not alone. Kent said, "People too often dismiss horror, and they dismiss the power of a scary film. I think it's a mistake to do that. Horror has the potential to be really profound and to tackle taboo subjects, and that's why I love the genre so much."[9]

An interesting filming aspect of *The Babadook* is that six-year-old child actor Noah Wiseman was shielded from certain aspects of the filmmaking. For example, during scenes in which Noah's character, Sam, is being verbally abused, an adult stand-in would take the place of Noah in reverse shots so he wouldn't be subjected to the yelling. This technique was also used by Stanley Kubrick when filming *The Shining* (1980) to protect the actor playing Danny Torrance (Danny Lloyd) from the more horrific elements of the film. Kent said, "Trying to direct any six-year-old is like trying to get an incredibly drunk person to perform in a straight line. It's not an easy task. But it was also an absolute joy." She added, "the thing about Noah is that he's a very lovable little kid. We auditioned quite a few kids who were very good, but I think they would have erred on the side of being really super annoying. A lot of that empathy is due to Noah—he's a strong actor."[10]

The Babadook itself was created very specifically using old-school special effect techniques inspired by Georges Méliès. "There can be something really visceral about things created in-camera," Kent said. "You get the feeling there was something there."[11] In-camera effects, stop motion, and puppetry were used to create the look and feel to the film. She also stated:

I felt like, for a creature like this to exist, the world itself would have to allow it to happen. So, I think if it was a naturalistic-looking world, and these things started happening, it would be quite ridiculous. So, it was important that the world of the film reflected the pop-up book at the center of it. We wanted a world that was heightened, but still felt like a real time and place.[12]

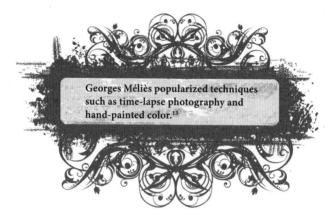

Georges Méliès popularized techniques such as time-lapse photography and hand-painted color.[13]

Kent also cites Edgar Allan Poe's *The Fall of the House of Usher* (1839) as an influence for the film. In 1928, French director Jean Epstein adapted the short story into a horror movie. Like most gothic stories from that era, *The Fall of the House of Usher* involves a decrepit mansion, people with health issues, and the general sense of impending doom. With the use of shadows and other visual elements, dread grows throughout the story. *The Babadook* makes similar use of these techniques, and as the mother starts to question reality, the house itself seems to become more of a character.

Unlike many horror movies, *The Babadook* doesn't end with the beast being eradicated. Amelia can't entirely get rid of her grief, but she is able to live with it by locking it in the basement. This feels like a satisfactory ending, especially for those who have suffered from mental illness. The condition, like the monster, may never be vanquished, but it may be able to be controlled.

CHAPTER THREE
BATES MOTEL

Years of Production: 2013–2017	
Developed by: Carlton Cuse, Kerry Ehrin & Anthony Cipriano	
Starring: Vera Farmiga, Freddie Highmore	
Network: A&E	

"A boy's best friend is his mother." When Norman Bates says this famous line in *Psycho* (1960), moviegoers cringed. On a subsequent viewing of the film, the line takes on new meaning and even more weight, and arguably more cringe. The television series *Bates Motel* (2013–2017) was able to explore that mother/son relationship even more than the trilogy of films was able to. Part of the brilliance of the series is that we know the ending: Norma Bates will become the putrid, rotting corpse that her son keeps in the house. But how do they get to that point?

The Wi-Fi password for the Bates Motel is MOTHER.

Bates Motel is a prequel to *Psycho* and was conceived by showrunners and producers Carlton Cuse and Kerry Ehrin. It was important to Ehrin to have Norma Bates, portrayed by Vera Farmiga, be shown as a real woman. They "collaborated closely to ground the character in reality, far away from either idealized motherhood or Norman's sick projections of her, by rooting the character in her own lived experiences."[1] The mother in the movie *Psycho* is only alluded to and referred to by Norman in the beginning. We catch glimpses of her silhouette and are given hints about her personality and temperament by Norman. She's ever-present because she is Norman, and Norman is her. The audience doesn't discover this until the end, of course. The mother is the monster that Norman, and our own imaginations, have painted her to be. Norma herself isn't a monster, though. Her greatest flaw may be her helicopter parenting and her inability to let her son make mistakes and decisions on his own.

Taxidermy began in England in the 19th century.

In reference to creating a character and backstory for the television series out of the source material, Ehrin said, "It's interesting in a world that has historically been defined by men, even on the page, to say, okay, this woman was blamed for everything that was wrong with this guy. Let's take a look at that. Let's take a look at who she was as a whole person." Farmiga added, "The north star of the story that we were going to tell was that it was a love story between a mother and son, and about that umbilical cord unraveling and eventually potentially severing."[2] It may be easy to blame the mother for her child's behavior. But how much does nurture play a role in a person's actions? This is a point that *Bates Motel* explored. Norma was trying to be the perfect mother. She did things right. She tried to protect Norman from the darkness that haunted his life, and she thought by sheltering him, he would be safe.

Norman keeps his mother's dead body in his house. In reality, Elmer J. McCurdy's embalmed dead body was put on display and on tour around the US for forty years.[3]

There is a nuanced, and sometimes not so subtle, feeling of attraction and sexuality between mother and son that is present during *Bates Motel.* Norman seems jealous of his mother having any other close relationships with men and especially those that are physical. In film and other media, the female body is sometimes seen as a place. Luce Irigaray's 1984 lecture, "An Ethics of Sexual Difference," describes the existing model of woman as an othered object: "It is understood that she accedes to generality through her husband and her child but only at the price of her singularity. She would have to give up her sensibility, the singularity of her desire, in order to enter into the immediately universal of her family duty. Woman would be wife and mother without desire."[4]

Mothers will often do anything to protect their children. Debra Messing portrayed a mother in *Searching* (2018) who was willing to use her position in law enforcement to manipulate a case and ultimately protect her child. Norma, even against her better judgment, ultimately only wants to protect Norman and keep him from harm. It's, in a way, easier to live in denial than to face the truth about him.

We spoke to Dee Wallace, who has famously portrayed mothers in both *E.T.* (1982) and *Cujo* (1983), about working in the film industry.

Meg: **"One reason we are drawn to your performances is your dynamism in the face of typical horror tropes. For example, Donna Trenton in *Cujo* is a complex, authentic woman with good traits and bad. She bucks the stereotype of an all-perfect and sacrificing mother. Did you feel that there were many roles like Donna in horror available to you and your female acting peers?"**

Dee Wallace: "I do, actually. From *The Birds* (1963) to *A Quiet Place* (2018), strong female roles in this genre abound."

Kelly: "**You have devoted a lot of your recent work to promoting the very important concepts of self-love and self-esteem. Do you think horror media can be a positive space for young women? If so, how have you seen this manifest? And has it improved over the years?**"

Dee Wallace: "I'm not sure I would go so far as to say it is a positive space for young women. It certainly is for young women actors. We have moved from these roles being total victims, to real women strong enough to conquer the fear and the fate. In this way it has certainly improved."

Meg: "**Whether in horror projects you've been involved with, or ones you have enjoyed, have there been female role models (characters or creators) who stand out to you? Ones who promote female agency, confidence, or authenticity?**"

Dee Wallace: "So many. The new female directors are taking the industry as a whole by storm. Katherine Hepburn in the past was *the* quintessential strong woman in film. Bette Davis, Barbara Stanwick. Then we went into the 80s victims, and now we are back to really strong, vital, smart women both in film and TV."

Kelly: "**Is there a misnomer about the horror industry you often find yourself correcting?**"

Dee Wallace: "Always. The Academy and industry as a whole still do not consider the horror genre and the actors therein worthy of awards. It demands some of the highest emotional work in our industry."

Meg: "**In many ways, females created the horror/gothic genre (thank you, Mary Shelley!). Why do you think women are drawn to watch, create, and be a part of the darker side?**"

Dee Wallace: "I think everyone is drawn to it. It allows us to experience our fears in a safe place. As actors, it allows us to really show what we can do. And let's face it, it is just more of a ride to see someone strong like Sigourney Weaver win out over the bad alien than some muscle guy with a gun."

Kelly: "**You've played a mother in several of your horror movies. How did you approach working with children in your roles in *Cujo* and *E.T.*?**"

Dee Wallace: "I didn't have children when I shot those movies, but I had a really strong mother and grandmother that were my life's models for any mother I would play in my career. As far as working with kids, I treated them with respect and also a lot of TLC on the side."

Meg: "**Being a mother yourself, were there any expectations about motherhood that didn't turn out the way that you expected?**"

Dee Wallace: "It is my grandest role and my greatest accomplishment. And yes, much of it didn't turn out the way I expected. It's better."

Kelly: "**Tell us about your radio show, current projects, and where we can see you next!**"

Dee Wallace: "My radio show is a free call-in show where callers can ask questions and my channel answers. It is all about love, taking charge of your life, and being the creation of yourself through choice, focus, intention, and action."

A mother who surpasses all expectations of what a loving, maternal figure should be, yet subverts them within the genre of horror, is Morticia Addams. With her creepy, gothic look, dark sense of humor, and aversion to anything outwardly loving, Morticia doesn't seem to be a woman who would be up for any "Mother of the Year" awards. Yet, numerous examples of her character throughout her many incarnations prove that she is loyal, open-minded, and loving beyond measure.

Conceived by Charles Addams, The Addams Family became a household name in 1938 with its single-panel comics published in *The New Yorker*. Started as a humorous panel, the family became a part of American culture, and they have been featured in a live action television series, movies, and a new animated feature in 2019. The family shares macabre interests and dark humor but are not evil. Morticia and Gomez are shown as having a loving, healthy marriage including an active and consensual sex life. This was relatively unheard of on television shows of the 1960s, which usually consisted of showing husbands and wives sleeping in separate beds and humor sometimes being bred from a place of irritation instead of devotion. We haven't yet seen how Wednesday and Pugsley grow up in film adaptations of the family, but we can assume they are stable and contributing members of society, unlike Norman Bates. As television critic Emily VanDerWerff said, "The horror of motherhood is that you can know everything about your child, can be certain of the right path for him, and still be undone by the simple fact that monsters aren't created in momentous explosions. They're created by long paths lined with the best of loving intentions."[5]

Mothers will always be portrayed in horror movies and television shows as long as the genre exists. It's our hope that they continue to be shown in the plethora of complicated and interesting ways that explore the rich world of what it means to be a woman.

SECTION TWO
THE FINAL GIRL

CHAPTER FOUR
A NIGHTMARE ON ELM STREET

Year of Release: 1984	
Director & Writer: Wes Craven	
Starring: Heather Langenkamp, Amanda Wyss	
Budget: $1.8 million	
Box Office: $25.5 million	

A woman is being stalked by a killer in a horror movie. She is chaste. She is smart. She is able to outwit the villain and survives until the end. She is the final girl. We've all heard a character be described as the "final girl" in horror movies, but what does it mean? The term was coined by Carol J. Clover in her 1992 book, *Men, Women, and Chainsaws: Gender in the Modern Horror Film*, and is described as "typically sexually unavailable or virginal and avoids the vices of the victims like illegal drug use. Final girl is the 'investigating consciousness' of the film, exhibiting qualities of intelligence, curiosity, and vigilance; qualities that are traditionally associated with . . . male action heroes."[1]

To understand this concept, we need to first look at the idea of the "male gaze." Laura Mulvey describes it as "the act of depicting women and the world from a masculine, heterosexual perspective that presents and represents women as sexual objects for the pleasure of the male viewer."[2] For example, a movie scene featuring a woman will often view her as a sexual object. The camera may focus on her body and pan over it while she is in a passive situation as we view her from a male character's point of view. While many theorists label horror films as a male-driven or male-centered genre, Clover points out that in most horror movies, the audience is structurally forced to identify with the resourceful young female who survives the villain. While the killer's point of view may be male within the narrative, the male viewer is rooting for the final girl to overcome.[3]

The blades on Freddy Krueger's glove are meant to represent primal man, the claws of an animal, and our own primal fears.

Final girls existed before the term was used, of course. Lila Crane (Vera Miles) emerged as one of the first when she survived a harrowing confrontation with Norman Bates's "mother" in *Psycho* (1960). In *The Texas Chain Saw Massacre* (1974), Sally Hardesty (Marilyn Burns) is the only one to make it out alive from Leatherface's lair. Heather Langenkamp plays Nancy Thompson in *A Nightmare on Elm Street* (1984) and is a resourceful final girl who outwits Freddy Krueger (Robert Englund). As John Muir says in *Horror Films of the 1980s*, "Nancy is a rarity in the horror genre: an intelligent and insightful youth who is capable of connecting the important things in her life. Only Nancy can recognize the link between worlds for what it is, and look below the surface of reality because she is already trained to do so, through family history . . . Nancy is armed for battle and ready to rock."[4]

Nancy doesn't accept the fate of dying by Freddy Krueger's hand. Instead, she investigates his origins and comes up with a plan to fight

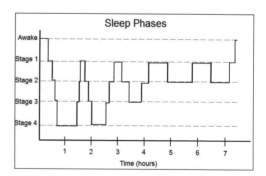

Nightmares tend to occur during the last third of the night when REM sleep is the strongest.[5]

back. She creates booby traps around her house and is able to pull Freddy out of her dream into the real world. After discovering he is fed by fear, Nancy calmly confronts him and makes him disappear as he tries to come for her. Freddy isn't defeated forever, of course, as the subsequent sequels prove, but Nancy emerges as the relatable hero of the film.

Langenkamp was cast by writer and director Wes Craven because he said, "She did not look like your typical Hollywood ingenue. She had a great strength and honesty to her face."[6] Langenkamp has embraced the legacy of the films, and her character, by attending horror conventions all over the world and by meeting fans of the franchise. Stories are shared in her documentary, *I Am Nancy* (2011), about how fans have been inspired by the character to overcome their own personal demons. Langenkamp said:

> Everybody thinks they've heard all about the final girl, but it's really not a part of Wes's origin of Nancy. The cool thing is that people take something that has a lot of depth and they're able to put their own statement about it. And that's a lot of what we liked to explore: how each person takes what they see on screen and then retranslates it to their own lives. It's definitely give-and-take between the creators of the movies and the fans. And you see that in this film.[7]

The concept of the final girl has changed over the decades. Movies that feature this changed final girl include *Scream* (1996), *You're Next* (2011), and *Evil Dead* (2013). In *Scream*, Sidney Prescott (Neve Campbell) is not a typical final girl. The movie is considered:

> The first film to consciously subvert the slasher genre. The smart referential tone of *Scream* being a slasher movie breaks the fourth wall of film; scenes of final girl Sidney Prescott undressing and making love to her boyfriend are intercut with her friend Randy [Jamie Kennedy] telling viewers explicitly what not to do during a slasher movie in order to survive."[8]

Sidney is seen as the first postmodern final girl, and moviegoers were treated to metajokes and moments throughout the film. Audiences embraced this self-aware approach to horror, reminiscent of Wes Craven's *New Nightmare* (1994), in which characters are aware of horror movies and tropes within the genre. *Scream* ended up making $173 million at the box office and secured its place in horror movie history.

You're Next follows the story of a group of people being stalked by masked intruders during a family gathering. Right off the bat in the film we see a couple having sex. Following typical form, they die a bloody, gruesome death. But that's where the clichés end. This isn't a film with a virginal final girl. Ehrin (Sharni Vinson) is a tough, no-nonsense kind of person and gets straight to solving the problems the group is faced with. A communication theory called the Johari Window states that we all have four selves: open, hidden, blind, and unknown. The unknown self contains all of the things we don't know about ourselves. Horror movies are full of characters discovering their unknown selves. They, and we, don't know how we'd react to certain situations until we're faced with them. Could we fight off faceless assailants? Would we be the first to die? Would we survive to be the final girl? We don't know until we, heaven forbid, experience it.

The character of Ehrin has a hidden self. These are the things she knows about herself but no one else does. Her boyfriend, played by AJ Bowen, even says, "I've never seen you act like this before." She responds, "It's a unique situation." Ehrin gives orders, is coolheaded, and keeps the group organized. These traits aren't always found in women in horror movies. Oftentimes they are like the mother, played by horror icon Barbara Crampton, overcome with grief and terror, sobbing uncontrollably. When given the chance, Ehrin takes out all three of the bad guys in a calculated (albeit violent) way. We find out that she grew up learning about being a survivalist from her father. She's been preparing for what-if scenarios her whole life. Ehrin defeats the other two villains and comes out victorious, though not unscathed. The last moment of the film shows us Ehrin, confronting her crooked boyfriend, listening to a speech about all of the things she could have if she plays along with his nefarious plot. She is our final girl and kills him, but not without being shot herself by the police. It's reminiscent of *Night of the Living Dead* (1963), although she is only shot in the arm and presumably survives.

The reboot of *Evil Dead* (2013) features a female protagonist. Unlike other gender-swapped films that received a lot of negative backlash, *Evil Dead* somehow subtly slipped past the angry fanboys by introducing us to Mia (Jane Levy). She's our final girl and our hero in the film, although that, too, is revealed slowly. The original *The Evil Dead* (1981) followed Ash Williams (Bruce Campbell) and his friends as they stayed at a cabin in the woods. The reboot sets up the audience with an almost identical plot and even teases us with a character who could be the new Ash: David, portrayed by Shiloh Fernandez. Mia survives and is shown as a flawed, realistic character. She's a recovering drug addict and becomes possessed by a demon in the film, just as the sister character did in the original. The movie goes against our expectations, though, and our final girl is the character we, perhaps, didn't think would survive.

The 2013 filming of *Evil Dead* reportedly used 70,000 gallons of fake blood.[9]

The notion of the kick-ass final girl, sometimes referred to as the "Amazonian" final girl, is an athletic, tough, and quick-thinking leader. We've seen her in movies like *Alien* (1979) and *Resident Evil* (2002). Women are getting more physical, action roles in film and television and within the horror genre specifically. We spoke with actress and stunt person Jenna Kanell to learn more about her experience.

Meg: **"You're an actress and performed stunts in movies like *The Bye Bye Man* (2017) and *Terrifier (2016)*. Can you tell us how you got into stunt performing?"**

Jenna Kanell: "When I was young, my dad enrolled me in karate classes to make himself feel better about my safety. I practiced Ho Shin Do, Tae Kwon Do, and a little bit of soft style Tai Chi. As an adult, I started training in Krav Maga (which I still do regularly).

Meg: **"My son takes Tae Kwon Do and he'll be excited to hear he could have a future in stunts!"**

Jenna Kanell: "I'd always thought that stunts were jumping out of buildings, running around on fire, what have you. When someone pointed out to me that this sort of athleticism was actually a stunt in itself, I did some research: turns out even holding a weapon is typically considered a stunt. Same with running. So, I took a couple of stunt-specific boot camps and began marketing myself as a union actor who did my own stunts. I've now had the honor of fighting a variety of foes and dying a number of times."

Kelly: **"Coming from a theater background, I've seen weeks of rehearsal go into a single stunt sequence and then each night it's rehearsed before the performance. What is the process like to create a stunt sequence for film?"**

Jenna Kanell: "It depends on the project's budget. Typically, the stunt coordinator choreographs everything and shoots a 'pre-viz' illustrating the action and types of shots needed to achieve it. They then present this to the director for approval. Once the powers that be sign off on it, rehearsals begin. These usually take place in a gym until the location becomes available for blocking, a walk-through, and on-set rehearsals involving the other departments. Then the action begins!"

Meg: **"Tell us about your film *Bumblebees* (2015)."**

Jenna Kanell: "Our autistic romantic comedy *Bumblebees* was the first short film I wrote and directed and stars my younger brother Vance. I essentially wanted to illustrate that regardless of physical and cognitive abilities, most of us just want to be loved and understood. I tired of watching exceedingly tragic films about people like Vance, treating autism like a death sentence rather than one facet of an individual as layered as any other. We ended up taking the film on not only a festival tour, but an educational run, speaking at institutions such as Harvard Medical School and the National Inclusion Institute. I even had the privilege of delivering a TEDx Talk (available on YouTube) on Vance's story and the concept of limitations. The mission was always to normalize the subject matter by facilitating dialogue on neurodiversity and compassion."

Kelly: **"We definitely need more representation like that!"**

Many women have never identified with the strong, kick-ass final girl character. A meme circulating the Internet pictures Negan from *The Walking Dead* (2010–present) and Harley Quinn from *Suicide Squad* (2016), both brandishing their signature bats. "Some people look cool holding a bat," reads the meme. "And then there's Wendy." Pictured below this text is Shelley Duvall, bat limp in her hand as she is menaced by her fictional husband in Kubrick's version of *The Shining* (1980). It's funny, harmless really. Yet, it's bothersome. Wendy Torrance, in both the novel and the film (and there are vast differences), is a real person. She is authentically navigating through her husband's alcoholism. She is a mother, scared for her child who is displaying disturbing abilities. She is stretched thin with worry, and then pure horror. She is not styled up like Harley Quinn, and she certainly doesn't hold the maniac authority of Negan. Wendy is real. She, and this is the pill no one wants to swallow, is all of us. She screams, she makes some poor choices, and her complicated past predicts her future. Allowing female characters to reveal the truths of their faults is a rebellion against the inconsistency of how males and females are often represented in media. Horror is a place of a high stakes. It is rife with possibility for complicated, relatable female characters to burgeon. Wendy may not look menacing in her corduroy jumpsuit, holding that bat with a trembling chin, but she's the hero of her own story. Not every final girl is, or needs to be, physically strong to survive until the end of a horror movie. Having a variety of women represented, in all genres, is better for all viewers.

CHAPTER FIVE

TALES FROM THE CRYPT: DEMON KNIGHT

Year of Release: 1995	
Director: Ernest Dickerson	
Writers: Mark Bishop, Ethan Reiff & Cyrus Voris	
Starring: Jada Pinkett Smith, Billy Zane	
Budget: $12 million	
Box Office: $21.1 million	

Horror movies in the 1980s gave us a multitude of final girls including Laurie Strode (Jamie Lee Curtis) in *Halloween* (1978), Kirsty Cotton (Ashley Laurence) in *Hellraiser* (1987), and Alice Hardy (Adrienne King) from *Friday the 13th* (1980). They all, like most final girls of the decade, had one thing in common: they were white. The 1990s saw a shift with one movie in particular, *Tales from the Crypt: Demon Knight* (1995).

Jadie David is the first African American woman to make a living in Hollywood as a stunt performer starting in the 1970s. She performed stunts in over forty movies including *Dr. Black, Mr. Hyde* (1976) and *Poltergeist III* (1988).[1]

Tales from the Crypt: Demon Knight is an offshoot of the television series *Tales from the Crypt*, which premiered on HBO in 1989 and ran a total of ninety-three episodes. Based on the original stories from EC Comics, the show featured anthology-type episodes focusing on all sorts of gruesome and gory plot lines. The movie version was written prior to the television series but wasn't made until 1995. "I've always enjoyed films that did a bit of genre-blending and I thought the story for *Demon Knight* did that perfectly," explained director Ernest Dickerson. "It was the perfect storm of horror, humor, thrills, had some mystery to it, and a lot of emotion, so I immediately knew this was going to be a very atypical horror movie to be involved with."[2]

Dickerson insisted on casting Jada Pinkett Smith as his female lead. The movie studio initially wanted Cameron Diaz for the role, but Dickerson put his foot down and gave audiences a strong, black final girl, saying:

> I was trying to figure out who can be my Jeryline. I always saw her as a small lady. I had gone to see *Menace II Society* [1993] and I saw Jada in that and I said, "That's her!" Now, Joel Silver, he wanted Cameron Diaz and I just couldn't see Cameron in that role. I keep pushing Jada and finally, he agreed to meet her. She said, "What do I say to him?" I said, "Just be yourself."[3]

The character of Jeryline isn't a typical final girl, and she isn't afraid to show it. She's got a sassy, bad attitude when first introduced and is a little rough around the edges. She's on probation for theft and is working at a hotel under the watchful eye of her employer. When demons attack, she helps organize the group to fight back and ultimately becomes the hero in the end. She singlehandedly outwits The Collector (Billy Zane) and uses her entire body as a weapon by covering it in the blood from the key. A perfect setup for a sequel that would never be, Jeryline takes over for Brayker (William Sadler) and is the protector of humankind. She becomes, presumably, the first woman in the long line of men who held this position. It's unexpected and exciting.

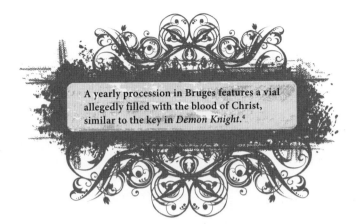

A yearly procession in Bruges features a vial allegedly filled with the blood of Christ, similar to the key in *Demon Knight*.[4]

Black women had been portrayed in the horror genre for decades but often only within certain tropes such as the magical "voodoo woman," characters who were disposable or sacrificial, or who served as a sidekick to the main character. As Harry Benshoff wrote:

> The American horror film often hinges on filmically constructed fears of the Other, an Otherness both drawn from and constitutive of any given era's cultural history. As many theorists have pointed out, the generic pattern of the classical American horror film oscillates between the "normal," mostly represented by the white, middle-class heterosexuality of the film's heroes and heroines, and the "monstrous," frequently colored by racial, sexual, class, or other ideological markers.[5]

Some films and theater productions became groundbreaking, although still using some of the pervasive themes, such as Orson Welles's production of *Macbeth* in 1936, which featured an all-black cast. The witchcraft in Scotland was replaced with voodoo in the Caribbean for the play and became a box office hit for the New York theater community.

The 1970s birthed several horror movies that featured black women in prominent roles including *The Omega Man* (1971), *Blacula* (1972), and *Abby* (1974). These "blaxploitation" films of the 70s had a major influence on American cinema. This genre is defined by having African Americans as the heroes and subjects of the film, rather than being portrayed as sidekicks or villains. The 1980s saw a handful of black women in horror movies like *Breeders* and *Vamp* in 1986.

Demon Knight (1995) featured several female characters including CCH Pounder as Irene, the owner of the establishment the film is set in, and Brenda Bakke as Cordelia, the sex worker. To Bakke, Cordelia was more complicated than the standard trope of sex workers that audiences see so often in the movies:

Macbeth opened on April 14th, 1936, and featured an all-black cast. Edna Thomas is pictured as Lady Macbeth.

> I never saw Cordelia as a "whore;" she was damaged and a little sad and pathetic too but there was a lot going on with her that wasn't written out on the page and that intrigued me. You don't see that too often with horror scripts. I just thought she had some very interesting character dynamics, especially with Roach [Thomas Haden Church] and Wally [Charles Fleischer], and I liked the fact that despite all her flaws, she still had this sweet personality to her which made her likable.[6]

Pounder, as the character of Irene, is tough and continues to fight back even after her arm is ripped off by a ghoul. In one scene, she lifts her bloody stump in the direction of The Collector and says, "That's me giving you the finger..." Although she dies a heroic death a few minutes later, Irene ultimately falls into the trope of the sacrificial black character.

We spoke to Ashlee Blackwell, creator of the site Graveyard Shift Sisters and writer and producer of the documentary *Horror Noire* (2019), to learn more about the representation of black women in horror. Blackwell's site is "dedicated to the scholarship surrounding the experiences, representations, achievements, and creative works of black women and women of color in the horror and science fiction genres."

Kelly: **"Why did you start Graveyard Shift Sisters?"**

Ashlee Blackwell: "I've always been a lifelong horror fan, and I noticed there was a lack of people discussing and delving into women of color [in horror]. With Dr. Coleman's book *Horror Noire* (2011), and another woman [Kristina Leath-Malin] who was in college at the time who was interviewing her for her own master's thesis project called 'My final girl: Black Women in American Horror,' these people together were doing some really creative work that had never been done before and was never discussed in the forefront of horror and horror journalism. One day I got kind of fed up with this lack of coverage and lack of discussion about women of color in the genre and decided to start the blog."

Meg: **"Can you tell us about how *Horror Noire* came about?"**

Ashlee Blackwell: "Phil Nobile Jr., who's now the editor of *Fangoria*, was working for a local production company and had been reading Graveyard Shift Sisters for a long time, and we finally met in person. He had a general understanding from the research that I've done that there was a whole history before what *Get Out* (2017) did as far as breaking ground for us finally discussing black participation in horror. Now everyone's talking about it, and it wasn't that way before. So, he brought me on board asking me to produce this and put it together."

Kelly: **"A theme I noticed in *Horror Noire* is that several roles for black actors weren't written with race in mind but the actor was the best choice for the role, which ended up bringing more nuance or impact to the film. Can you talk a little about that?"**

Ashlee Blackwell: "Even though sometimes certain roles are written 'race neutral' so to speak, once you put a person of color in that role, it definitely adds more to it. I think that's why audience reception and film criticism in general is so important because we're not just fans, we're critical thinkers. We're looking at it in different ways, and I think for women of color, and black women specifically, we're looking at these particular roles as something very revolutionary. We don't, every day, get a black final girl, so those characters mean more even though they may not have been originally written to have those deeper meanings."

Meg: "**Do you remember seeing *Demon Knight* for the first time? What were your initial thoughts?**"

Ashlee Blackwell: "It did stand out because she [Jerilyn] was the sole survivor, and then I started thinking about it even more. Look at her from the perspective of Dr. Coleman's research on 'The Enduring Woman' and also Carol Clover's 'final girl.' Those things combined make a really interesting perspective when you look at *Demon Knight*. That's what I like about it. Her being black gives it another dimension, and it's not just about her survival; it's about how black women consistently have to endure oppressive forces as it continues on. I think that's one of the most unique aspects of *Demon Knight*."

Kelly: "**What TV shows do you recommend for representation of women of color in horror?**"

Ashlee Blackwell: "I think *The Passage* (2019) was really good and hopefully gets a season two. *Z Nation* (2014–2018) unapologetically put a black woman in the lead role from the very beginning, and it kept that continuing on for the entire series. *Channel Zero* (2016–2018) did a lot with their last season. You saw this gradual push for women of color on-screen and particularly black women."

Meg: "**Where do you see room for growth?**"

Ashlee Blackwell: "Honestly just giving the opportunities for filmmakers who want to tell their own stories, who are black women, and have doors open for them. I don't see that a lot, unfortunately. I want to see multiple women, who have multiple perspectives in the genre, tell their stories. Rachel True says, 'Let us tell our stories. You'll see things you've never seen before.' I really, truly believe that."

Kelly: "**Absolutely!**"

Another underrepresented group within the horror genre is indigenous women. The first native female director of a feature-length film is Georgina Lightning, who also wrote, produced, and starred in the horror thriller *Older Than America* (2008). The movie tackles the intergenerational impact of the forced attendance of boarding schools on indigenous children. Lightning noted that "over 50 percent of the students who attended boarding schools died in boarding schools . . . and even though 1975 was the last account of mandatory assimilation, the transgenerational trauma continues to affect everybody."[7] For more about intergenerational trauma, see the next chapter, on 2018's *Halloween* (p. 40). The TV show *Chambers* (2019) features the first indigenous female lead on a Netflix series. Sivan Alyra Rose, who portrays the character Sasha on the show, said, "I don't wanna be the token native. I want the industry to see me and go, 'Oh my God, we should go to the res to find kids' and I hope it's just a good spark of representation in the big pot of Hollywood."[8] We still have a long way to go with representing women of color in the horror genre, but these small steps will surely lead to larger ones.

CHAPTER SIX
HALLOWEEN

Year of Release: 2018	
Director: David Gordan Green	
Writer: Danny McBride	
Starring: Jamie Lee Curtis, Judy Greer	
Budget: $10 million	
Box Office: $253.7 million	

If you've seen the original *Halloween* (1978), there are many moments that may resonate with you: the stark, haunting music; the silent, faceless killer; or the unrelenting force of Laurie Strode (Jamie Lee Curtis). One aspect that stuck out to many young people after watching the movie is the concept of the babysitter. It's a position most young girls hold in their lifetime, and seeing it portrayed on screen in a terrifying but realistic way made many of us question the noises we heard while being responsible for younger children. My coauthor, Meg, remembers babysitting for the first time when she was twelve. I (Kelly) began babysitting with my older sister and undoubtedly thought of Michael Myers lurking outside the door. Thankfully, babysitting for us turned out to be generally uneventful besides—cleaning a few poopy diapers!

Babysitters being stalked is a common trope in horror.

The term "babysitter" began to be used in the United States in the 1930s. The position had always existed, of course, but was referred to as child-rearing, nursing, or nannying. Those who needed childcare often had family close by who could watch their children or had servants or nannies already in their employ. It was the baby boom post-World War II that led to an increased need for babysitters in the suburbs. Teenage girls, and sometimes boys, were looking for employment, and it became the main moneymaker for this demographic.[1]

There are several urban legends, and subsequent films, that delve into the trope of a killer stalking a babysitter and her charges. Perhaps the most famous is the story of the babysitter and the man upstairs. A young girl is bothered repeatedly by phone calls telling her to check on the children upstairs. After getting the police to trace the calls, they warn her, "The call is coming from inside the house!" This was seen in *Black Christmas* (1974) and *When a Stranger Calls* (1979). This urban legend may have been based on a real-life occurrence. In 1950, a thirteen-year-old named Janett Christman was babysitting for a family in Columbia, Missouri. The local police station received a call at around 10:30 p.m. from a young woman "screaming in sheer panic" to "come quick!"[2] Long before caller ID, the police were unable to trace where the call was coming from. It wasn't until after 1:30 a.m. that the couple returned to find Christman murdered on their living room floor. Their three-year-old son was upstairs sleeping, unharmed.

It was this real-life experience of babysitting that inspired cowriter Debra Hill to include it in the plot of *Halloween*. Laurie Strode is babysitting Tommy Doyle (Brian Andrews) when she continues to be stalked by Michael Myers (Nick Castle):

> It was Hill's secondary voice on the script that ended up being vital to the success and durability of the film . . . her contribution to the dialogue and the relationships between the female characters that sets the film apart even by contemporary standards. The rapport between the teenage girls and how they interact with those around them feels organic and believable in a way that can only have been written by someone who had had similar experiences.[3]

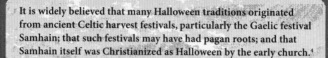

It is widely believed that many Halloween traditions originated from ancient Celtic harvest festivals, particularly the Gaelic festival Samhain; that such festivals may have had pagan roots; and that Samhain itself was Christianized as Halloween by the early church.[4]

Laurie Strode escapes Michael Myers in the first film, and the 2018 sequel picks up forty years later. Jamie Lee Curtis acknowledged that the character's experience would be shaped around the trauma and posttraumatic stress that she experienced:

> I imagine she went back to high school two days later. This happened on October 31st and maybe her parents let her stay home from school on the first and second of November but by the third of November she was back in school. I'm sure she just became a freak, meaning she's the girl that survived but she was already a vulnerable loner and I'm sure she just became more so. Of course, all the trauma was just buried and I just don't think she ever got any help so I liked that the movie explored what actually happened to Laurie Strode for all those years.[5]

Post-Traumatic Stress Disorder (PTSD) is a real mental health condition that was often referred to as "shell shock" in past eras. Anyone going through a traumatic experience can suffer from it, and it afflicts over three million people in the United States alone.[6] Symptoms include intrusive thoughts and memories, upsetting dreams or nightmares, and severe emotional and physical distress. These symptoms can occur almost immediately following a traumatic event or can take years to surface. Laurie would most certainly have changed her approach to life based on her experience of the one night she was stalked by a killer.

Laurie becomes an untraditional final girl in the 2018 film. First of all, she is imbued with knowledge about her situation and potential killer. Having lived through the experience before, she has had time to prepare and plan for worst-case scenarios. Second, she is an older woman. We don't often see final girls in their fifties or sixties. Her life has been lived and based on the trauma she dealt with. Her entire being has led her to this time and place to become the final girl in her own tale once again. Speaking of age, *Halloween* in 2018 became the biggest opening for a film with a woman lead over the age of fifty-five in history.

In the 1978 film, Laurie is a more typical damsel-in-distress. She reacts, successfully, to escape her potential killer but doesn't outwit or outfight him in the end. Dr. Loomis (Donald Pleasence) ultimately saves her by shooting Michael Myers. We often refer to this act on our *Horror Rewind* podcast as "stealing the climax" from the female lead. The woman fights until the very end but is ultimately saved by a man.

Speaking about the 2018 film, Director David Gordon Green said: "There have been so many clichés or tropes, or however you want to look at it. We try to embrace them and then turn it upside down a little bit. So, if your expectations of a final girl are of some kind of damsel-in-distress, your not going to get that in this movie."[7] As Jamie Lee Curtis said, "Trauma is generational and if it's untreated, it is life-changing until you can take back the narrative." This is exactly what happens in the film. Laurie is able to become the rescuer and protector in her own story through her resiliency and strength. Transgenerational trauma can affect people more than they realize. The condition was first observed in the 1960s in a group of children of Holocaust survivors but roots further back to African American families and the trauma experienced from slavery and racial discrimination. The lasting effects of a traumatic event can remain and impact descendants for generations. Building upon the clinical observations by child trauma researchers, experts have "empirically identified psychological mechanisms that favor intergenerational transmission . . . as an effect of parental efforts to maintain self-regulation in the context of post-traumatic stress disorder . . ."[8]

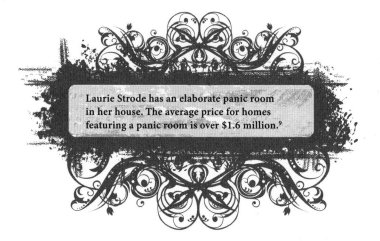

Laurie Strode has an elaborate panic room in her house. The average price for homes featuring a panic room is over $1.6 million.[9]

This trauma in *Halloween* undoubtedly affected Laurie's life and her family's lives. When we learn more about what her character has been up to over the past forty years, we learn she is twice divorced, has a strained relationship with her daughter and granddaughter, and has devoted her life to weapons training and building her compound, complete with safe rooms and traps. The film also touches on a topic not often explored: the link between victim and assailant. Although the sibling plot of the *Halloween* sequel is dismissed in the 2018 film, there is still a connection between Laurie Strode and Michael Myers. He is her stalker, attacker, and almost killer. She is the one who got away. We may never know, or understand, what drives Michael Myers to kill. Arguably, it's almost scarier if we're unable to deduce the psychological or scientific reasons behind his drive. Random acts of violence or murder are the true horror of this world and can't be explained. Laurie, though affected, will not be defined by her victimhood but instead by her resilience and determination to never let history repeat itself. The final girl has become the final woman, and we can learn this important lesson from Laurie Strode: even when obstacles seem nearly insurmountable, we can prepare and fight like hell to overcome.

SECTION THREE
SEX

CHAPTER SEVEN
SLEEPAWAY CAMP

Year of Release: 1983	
Director: Robert Hiltzik	
Writer: Robert Hiltzik	
Starring: Felissa Rose, Jonathan Tiersten	
Budget: $350,000	
Box Office: $11 million	

I (Meg) never attended a sleepaway camp, but Kelly spent eleven summers at one in northern Minnesota. While there one year, she convinced her bunkmate that Freddy Krueger himself was outside their cabin window, smiling maniacally. Camp can be both fun and scary—leave it to an eight-year-old horror fan to focus on the latter!

Many of the female horror characters we've previously discussed in this book have been depicted as heteronormative. Because their sexual appeal is often played up on screen for the male gaze, their sexuality is depicted as aggressively "straight." But, what of the other women in horror? Those who exist somewhere else on the Kinsey scale, whether bisexual, lesbian, or somewhere in-between? We were

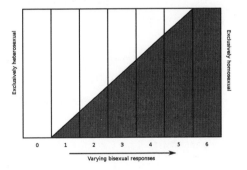

The Kinsey Scale was created in order to demonstrate that sexuality does not fit into two strict categories.[1]

also curious to learn about the representation of transgender women in horror. It was to our delight that we were able to discuss this important topic with horror filmmaker, and trans woman, Stacey Palmer.

Meg: "**First, can you tell us about your role as writer and producer? What sort of stories do you work to tell?**"

Stacey Palmer: "The two roles are a ying-yang sort of experience. With writing it's solitary. It's a relaxing experience. For the first draft or two anyway. Then I start to show it to people. For me there's a natural anxiety that comes. Of course, you want it to connect with people, no matter the genre. I work mainly in horror. I think there's a lot of subtle reasons why, but the one I recognize that drives me is the need to understand. In our world what causes so much unchecked evil? Where does that come from? My role as a writer is to explore that evil from an outside perspective. I am a fly on the wall examining the actions and motivations of the characters. I then have the responsibility to dictate them the best I can. There is a level of control I have on the story. Many times, though, the story will take me in a wild direction. I just follow it the best I can. Producing is the exact opposite. Producing is very strict. You are now responsible to make that story come alive. Put the skin over the soul, for lack of better words. Then you have to remember each body is controlled by many working variables. You are the brain and your crew is the lifeblood. You two need to work smoothly or you're dead."

Kelly: "**In a previous interview you confessed your love for *The Twilight Zone* (1959–1964)! What is your favorite episode? And how did it inspire your creativity?**"

Stacey Palmer: "Yes! The original *Twilight Zone* is my favorite everything! The writing, the performances, even the effects and makeup work for the world they exist in. I first discovered the show when I was ten. I was home from school sick in the living room watching game shows. We had no remote-controlled TV, so I was at the mercy of the network gods. Luckily, they took pity on me because after *Card Sharks* (1978–present) ended I heard those famous first notes of *Twilight Zone*'s theme song. I was half-asleep and a little annoyed that a black-and-white show was coming on. The episode started. I became entranced."

Meg: "I, too, am in love with Twilight Zone. It has also influenced my work, and it really stretched my imagination and understanding of the world."

Stacey Palmer: "The episode was "The Monsters are Due on Maple Street" (1960). It's about a small neighborhood thrown into hysteria because they believe one of them is an alien. The story then becomes a witch hunt with everyone turning on one another. It is a metaphor for the Cold War paranoia of the 1950s and 60s but was still very effective in the 80s and even today. It's an episode that taps into the fears of cultures that we don't understand, so we turn to anger and violence to help deal with that anxiety. This is the same sort of fear that the transgender community is facing now."

Meg: "Is there a trans character in horror that you feel was depicted with authenticity?"

Stacey Palmer: "As of today, I have not. Of course, I haven't seen every movie, so I will gladly accept being wrong."

Kelly: "How would you like to see trans representation improve in horror media?"

Stacey Palmer: "The problem with a lot of films that use transgender characters is that the film is created around them. The spotlight is shining on them. It's as if the producers are saying, 'Hey, look! We have a transgender character in our movie.' Instead they should be creating content and casting an actor for the part where they just happen to be transgender. You may never know because they are just another character. It doesn't matter if they're transgender or not. But they should never cast a cisgender actor in a transgender role. Here's why that becomes problematic: let's take *Dallas Buyer's Club* (2013) for an example. Jared Leto played a transgender woman. He even won an Oscar for the role. Now when he accepted the award, he was presenting male, because he's not a transgender woman. This is a problem for the trans community because now the world sees Leto in a suit and full beard. This sends the message that trans women are just men in dresses or they can turn it off and on when they want. That is not how being transgender works. Trans women are women, and we need more authentic visibility."

Meg: **"Please tell us about your current projects."**

Stacey Palmer: "Right now I'm working on a short body horror film called *Toothache*. I am also one of nine female directors working on a horror anthology based around Valentine's Day called *Love You to Pieces*. Past that, time will tell. Hopefully it will be long and rich."

Kelly: **"We hope so, too!"**

Sleepaway Camp is not a film about the transgender experience. The final reveal, that Angela (Felissa Rose) is indeed a boy, is the story of an abused, misgendered child who has no control or agency over their body. Their disturbed aunt makes them act and appear as a girl, but we have no reason or insight to believe Angela wants to live as a girl. That being said, the 1983 film, while on its surface is a teen slasher, delves into gender roles in a way very few films of its time did. On one hand, an empathy can be derived for the victimized Angela, a person who has been abused so severely they turn to murder. On the other hand, some film critics maintain that *Sleepaway Camp* displays a transmisogyny, including trans writer Willow Maclay:

> *Sleepaway Camp* is a curtain-yanking picture with a reveal that works only to make a woman with a penis a vessel for horror. It is a trap narrative rechristened in the structure of the slasher genre, where, instead of a lover being revealed as a deceiver, a murderer is exposed as a transgender woman. This harmful, false narrative furthers the notion that we are not who we say we are, and that we are more prone to committing violence than of being victims of it ourselves. The film asks its viewers to build sympathy toward Angela, who has had a difficult life leading up to this summer at camp, before transforming her into a monster in the final few

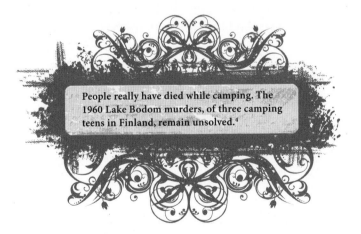

frames. By making Angela a whipping post for constant teasing she becomes the central character. In horror movie tropes she is set up to be the "final girl," someone you rally around when she is eventually confronted with the incarnation of evil. However, Hiltzik subverts this idea by making Angela not the victim but the killer, and—in what the movie suggests is a worse transgression—not a girl but a boy.[2]

Wherein lies the counterpoint, as some in the trans community have seen the positives in trans depiction in *Sleepaway Camp*. Nat Brehmer explains in "The Pros and Cons of Sleepaway Camp as a Trans Narrative":

It makes sense why there are people in the LGBTQ community who have embraced *Sleepaway Camp*. Not only is Angela not just another killer who happens to be a white male, she's a trans female antagonist and one of the only ones in horror history. At the same time, she is the protagonist of the feature. She's the main character. In this first film, she is a perineal outsider. Everyone hates her and most of them don't even know *why* they hate her.[3]

People really have died while camping. The 1960 Lake Bodom murders, of three camping teens in Finland, remain unsolved.[4]

Despite Angela's trans status, the character's complexity has transcended that of many others in horror. As with transgender characters, lesbian characters in horror films are often met with punishment, as lesbianism delineates from traditional expectations of the female. Like so many of the archetypes and tropes discussed in this book, lesbianism is seen as a rebellious act. In "Daughters of Darkness Lesbian Vampire Films," Bonnie Zimmerman argues that the subgenre of lesbian vampire films that peaked in popularity in the seventies coincides with the surge of women's rights: "The lesbian vampire, besides being a gothic fantasy archetype, can be used to express a fundamental male fear that women bonding will exclude men and threaten male supremacy. Lesbianism— love between women—must be vampirism; elements of violence, compulsion, hypnosis, paralysis, and the supernatural must be present."[5]

While the lesbian vampires of films like *The Vampire Lovers* (1970) and *The Velvet Vampire* (1971) may be exploited to strike fear in men's hearts, other queer women have been given authentic characterization. In *The Taking of Deborah Logan* (2014), Sarah Logan (Anne Ramsay) worries about her aging mother while still nursing feelings for her ex. There is also *Lizzie* (2018), in which Lizzie Borden (Chloe Sevigny) has a romantic affair with housemaid Bridget (Kristen Stewart) before the infamous "forty whacks." This supposition, that the two were lovers, is not completely unmerited, as the real Borden was known to be a lesbian. In the French film *High Tension* (2003), Marie (Cécile de France) harbors such love for Alex (Maïwenn) that she resorts to violent murder. Theo in all iterations of *The Haunting of Hill House* has been a queer woman, and Willow (Alyson Hannigan) and Tara (Amber Benson) were a groundbreaking pair in *Buffy the Vampire Slayer* (1996–2003).

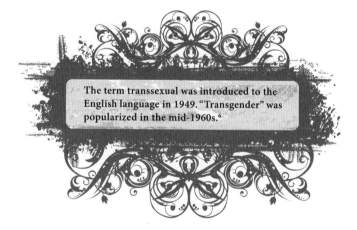

The term transsexual was introduced to the English language in 1949. "Transgender" was popularized in the mid-1960s.[6]

Although *The Rocky Horror Picture Show* (1975) is more camp and less horror, it has endured as a well-loved cult classic. Tim Curry's depiction of Dr. Frank N. Furter was ahead of its time in gender fluidity and exploration. We interviewed actress Alice Schroeder to learn more about her experience as a trans woman working in theater, including her turn as the famous "sweet transvestite from Transylvania."

Meg: "**Do you remember the first time you saw *The Rocky Horror Picture Show*? What impact did it have on you?**"

Alice Schroeder: "I remember it was my twelfth birthday and I could finally convince my parents to rent me the horror movies. I remember basing my choices off of the covers and what looked the coolest. Frank on the cover with the big red lips just seemed so naughty/intriguing even though I had no idea what it was about.

Kelly: "**I remember that cover at my local library! I watched it for the first time when I was twelve, too.**"

Alice Schroeder: "I planned a big birthday weekend and even invited a cute boy I had a major crush on. It was an overnight slumber party, and we binge-watched *Child's Play* (1988), *Halloween* (1978), *A Nightmare on Elm Street* (1984), and the last one was *The Rocky Horror Picture Show*. It blew my mind. Everyone was falling asleep, because it wasn't exactly scary . . . but I was just transfixed. It combined musicals and B-movie sci-fi horror into one. It was a dream come true. At that age I was coming to terms with my own sexuality/identity, and seeing a male actor portraying an eccentric transsexual with homosexual tendencies truly was a pinnacle moment for me. I realized that I wasn't the only one. I think that's what most children coming to grips with their sexuality need to see: representation. I always thought I was alone. To this day, the movie and stage show's pure wackiness and overall aesthetic has guided my creative drive and passion.

Kelly: "**You have played Dr. Frank N. Furter for several years in a stage production. Tell us about that experience!**"

Alice Schroeder: "Our shadow cast company has been together for going on nine years now. So, some of us have been in our character's heels for a long time. To me, Dr. Frank is the epitome of sex and freedom. He is everything that your parents tell you not to be. If you want to get real serious with it, you could easily compare him to Satan. Sitting on his throne, how they worship him and let him control their life, he is the nocturnal king/queen of the night. Men and women both want to fuck him. You can't buy that kind of influence.

Meg: **"What drives that character?"**

Alice Schroeder: "Pure power and self-indulgence. He is proud of his body. Nobody tells him no; he always gets his way. He doesn't know when to quit, and ultimately it ends up being his demise. Once you're in the part of Dr. Frank, what surprised me is how it is such a physical role. Physical slapstick humor while trying to be sexy in the most uncomfortable clothes. It's hard to keep a smug face when your feet and legs burn and makeup sweats into your eyes. Though the biggest challenge of performing in a shadow cast is you must somehow incorporate and mimic the pure genius of Tim Curry, who originated the role and made it what it is today. The confident, humorous, emotional, pure sexiness, and dedication given by Tim Curry makes Dr. Frank probably the most difficult horror villain to portray. People hate him and love him at the same time. You have to find that balance of good and evil within yourself."

Kelly: **"I love that! Horror, and all movies really, need that villain who has endearing qualities, too!"**

Meg: **"What have you learned from cast and audience feedback during your productions of *The Rocky Horror Picture Show*?"**

Alice Schroeder: "First and foremost *Rocky Horror* is about community. It's about having a family that is just as wacky, sexual, and strange as you. Transylvanians gathering together at the castle for a wild party. They are with their own kind; they stick together and I think that's what resonates the most with people. You'll see the same faces in the audience over and over throughout the years, bringing their wide-eyed virgin friends. We try our hardest to establish the theater as a safe space for all. It's about having fun and letting loose. We get feedback always saying how they've never seen a shadow cast perform in such a unique way, and that is something I'm extremely proud of. Even if it is a movie playing in the background, it is very much a theatrical experience (lights, projections, effects). It's quite a beast of a show. *Rocky Horror* shadow casts also provide opportunities for all types of artists. Visual artists and actors of all identities, shapes, sizes, and experience levels. It is a welcoming environment. As a cast, we grow close like family. We become intertwined in one another's lives. The friendships we create don't disappear as soon as the lights go off. We celebrate one another's weddings, pregnancies, we are there to console one another during breakups or tearful good-byes when people move away. It's a very special bond that is created by *Rocky Horror*, and you truly don't know how it feels until you're a part of it."

Kelly: **"Being a horror fan, how have you seen trans characters be portrayed in the genre? Do you think it's shifting?"**

Alice Schroeder: "As a proud trans woman, Dr. Frank is still very much an unhealthy stereotype of trans culture, but for the time it was a huge, *bold* statement saying that we exist. In horror, generally, cross-dressing and gay/trans characters are comic relief or mentally ill. It's a pretty old-fashioned concept. But, horror has always been purely entertainment to me. Do I see it being a main genre that will change the political climate toward certain issues? Absolutely not. But if used right, it can definitely be an outlet to provide opportunities for trans and queer artists and, most important, bring out issues and certain identities to be *seen* and represented."

Meg: "**Are there any characters, movies, or television shows that show good representation (in any genre, not just horror)?**"

Alice Schroeder: "I solely stand by my opinion that when it comes to portraying a trans/queer story, you must have a trans actor/writers/directors. There are emotions and experiences that a cisgender/privileged person can't harness and I find insulting to even try. I have to say as of right now in pop culture the FX show *Pose* (2018–present) is making a step in the right direction in providing accurate trans representation in entertainment. But what a dream it would be to have trans actors portraying roles written for cisgender actors . . . that's when we've truly changed."

It was our great pleasure to speak to both Stacey Palmer and Alice Schroeder about their experiences in horror. They are prime examples of women who are creating their own content and controlling their own narratives in order to further the horror genre.

CHAPTER EIGHT
TEETH

Year of Release: 2007	
Director: Mitchell Lichtenstein	
Writer: Mitchell Lichtenstein	
Starring: Jess Weixler, John Hensley	
Budget: $2 million	
Box Office: $2.3 million	

We were children when the infamous Lorena Bobbitt story hit the news. The headline became a punchline and people were aghast that a woman had cut off her husband's penis! Decades later, more of the story has come into public light and the atrocities that Lorena Bobbitt suffered at the hands of her husband have been revealed. How could a similar story be made into a subversive, tongue-in-cheek horror movie that is both darkly funny and feminist? The film *Teeth* (2007) does just that.

There is no question that often in horror films a woman's sexual agency is punished "immediately before a woman is brutally assaulted . . . she is seen disrobing, masturbating, engaging in sexual intercourse, sunbathing, etc. During and after her assault . . . these presumably mildly arousing, pleasing scenes are juxtaposed with the violent images."[1] Not only is the female character's consensual act punished, but exploited for the aforementioned male gaze, intercut with violence. Marion Crane (Janet Leigh) of 1960's *Psycho* is an example of a grown woman comfortable with her sexuality who pays dearly for this trait. When we first lay eyes on Marion, she has just made love with a married man, and we are viewing her through an almost voyeuristic gaze. Later, when she has booked a room in the world's worst motel, she is watched Peeping Tom-style by Norman Bates (Anthony Perkins) before she is murdered while naked in the shower. From a modern perspective, Marion's shower scene is downright Puritan, but by the standards of the time, her undressing is a sexualized scene. A sexualization that immediately turns violent. After all, this was the first time a flushing toilet was seen on screen!

Like Marion, a drove of horny teens of horror have met their end because they dared to seek out, and even, *gasp*, enjoy sex. The premise of the original *Friday the 13th* (1980) is a vengeful mother (Betsy Palmer) who lost her son to a drowning because camp counselors were getting frisky. When this son, Jason, is resurrected for the sequels, he finds sex as abhorrent as his mother did. He continually kills teens who are about to have, are in the middle of, or have just finished a sexual encounter. In the study outlined in "On the Perils of Living Dangerously in the Slasher Film: Gender Differences in the Association Between Sexual Activity and Survival," researchers chose fifty random horror movies to help them prove that females who engaged in sexual activity would be punished most severely in horror. This punishment would manifest in screen time of their abuse and death. First, the study found that women's death scenes were significantly longer than men's. Next, they discovered that there was no difference in the runtime of deaths of males who had not had sex versus those who did. As they guessed, female characters who had been shown engaging in nudity or sex had significantly longer deaths than those females who did not.[2] Their thesis was that women actively participating in their own sensuality were heartily punished by the monster, slasher, or whoever wielded the weapon. Because, as theorized before, the male identifies with the antagonist. On the other hand, Klaus Rieser provides a counterpoint of this interpretation in "Masculinity and Monstrosity: Characterization and Identification in the Slasher Film":

> . . . it may very well be, then, that the power and potency of the monster body in many classical horror films . . . should not be interpreted as an eruption of the normally repressed animal sexuality of the civilized male (the monster as double for the male viewer and characters in the film), but as the feared power and potency of a different kind of sexuality (the monster as double for the woman). Horror films have also been subject to historical changes: While the classical horror at least granted some power

to both the monster and the woman through their difference to the normal male, the post-*Psycho*, post-Peeping Tom films not only escalate the doses of violence, but . . . conflate the woman with the monster, often leaving the woman's body as the only site of horror.[3]

In *Teeth* (2007), this site of body horror exists within Dawn (Jess Weixler). A fiercely feminist film on women's sexuality, *Teeth* is about a teen inflicted with "vagina dentata," or a set of razor-sharp teeth in her vagina. While this is clearly a fiction, vagina dentata has

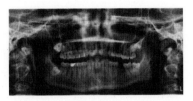

Vagina dentata is a myth found in several cultures.

existed in many cultures' folklore, sometimes to discourage rape. We see Dawn discover these myths through her search online. It becomes evident that the concept of a barbed vagina was also used to encourage men to "break" the woman, and that only a "hero" could vanquish these castrating teeth. As Dawn reads on, she finds that these primitive beliefs of men were in response to their fear of women's sexuality, and also of their own impotence.

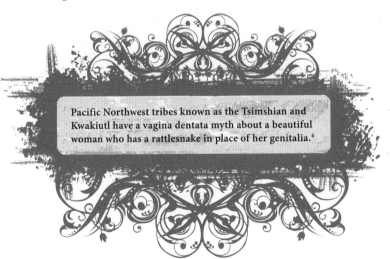

Pacific Northwest tribes known as the Tsimshian and Kwakiutl have a vagina dentata myth about a beautiful woman who has a rattlesnake in place of her genitalia.[4]

At the beginning of the film, Dawn is passionate about her virginity. She wears a purity ring and speaks publicly about the importance of chastity. The concept of the virgin is fundamental in understanding the characterization of women in horror. The very definition of the "final girl" includes the virginal.

Virginity itself is a myth, one that has been inculcated into every facet of media. In her book *The Purity Myth: How America's Obsession with Virginity is Hurting Young Women*, Jessica Valenti writes, "If we're to truly understand the purity myth, we have to recognize that this modernized virgin/whore dichotomy is not only leading young women to damage themselves by internalizing the double standard, but also contributing to a social and political climate that is increasingly antagonistic to women and our rights."[5]

This is on display when *Teeth*'s Dawn is in her high school biology class. The male teacher effortlessly uses the word "penis," yet when he asks the students to turn to the next page, he can't utter the forbidden word "vagina." To add insult, the anatomical drawing of the female reproductive organs is covered with a sticker. When the students point out that the male drawing is not obscured, the teacher explains the difference was made "for obvious reasons." Dawn, not yet awakened to her own sexuality, agrees with this systemized misogyny, saying women are just naturally more modest. *Teeth* is illuminating one of the central fallacies in our treatment of women both on and off screen: that men's sexuality is a comfortable, familiar place, and that women's sexuality (which seems to exist in the two extremes of whore and virgin) should be an unspoken desire, or "leave a little mystery to be desired." The film is showcasing that both men and women are prone to believe these antiquated ideals.

Women can be placed in the virgin and whore categories rather easily, with no concern to the multitudes of complexities in between. If the prototypical final girl exists in the sphere of virgin, most often her friends are given the whore mantle. As discussed in the teen innocence chapter (p. 95), the cheerleader trope often goes hand in hand with the judgment of whore. When virgin Laurie Strode is left alive at the end of *Halloween*, it's after her sexually active friends Lynda and Annie (P.J. Soles and Nancy Loomis) are killed. The purity of *Psycho*'s Marion Crane would certainly be called into question in 1960. It wouldn't be hard to imagine the word "whore" being bandied around because of her sexual choices.

Let us explore the actual connotations of the term whore. "Whore" means prostitute. And a prostitute is a woman who offers to hire her body for indiscriminate sexual intercourse, or so says *The Concise Oxford Dictionary*.⁶ Colloquially, whore has also come to mean a woman who has sex with whom she wants and unabashedly enjoys it. These, as we know, are not negative characteristics, yet the name and stigma persist. The conundrum that exists within this treatment of women as whores is that the male gaze both fetishizes and punishes. They both desire and are afraid of these often-adolescent objects of confusion. Women and girls, most often not the final girl, are shown in tight clothes. They are considered conventionally attractive and can be almost threatening in demeanor (think Tatum [Rose McGowan] in *Scream*). Speaking of *Scream*, we come to the subversion of the virgin trope in Sidney's (Neve Campbell) final girl. Sidney is "allowed" to have sex with her boyfriend (Skeet Ulrich) and still live. But, as Karen J. Reiner points out in her piece "Monstrous School Girls: Casual Sex in the Twenty-First Century Horror Film," this new rule is still troublesome:

> I argue that another equally problematic ideology has replaced this older one. Specifically, the new rule the teenage girl needs to survive a horror film is to only have "meaningful" sex, which usually means sex within the context of a relationship with a partner who is loving and loyal, equally (in)experienced, and conscientious about his partner's pleasure. By contrast, girls who engage in casual or "meaningless" sex are often killed off quickly in horror films, a plot device that communicates very straightforwardly that their behavior is forbidden.⁷

Next in *Teeth* comes Dawn's romance with Tobey (Hale Appleman). As they spend time together, it becomes increasingly difficult for them to stay true to their pledge of virginity. Dawn fights the natural urge within herself to have sex, believing in the above notions of purity. What comes next is a stark depiction of date rape. Tobey, who had previously seemed sweet, makes it clear that he believes Dawn *owes* him sex. He rapes Dawn, which leads to the discovery of her vagina dentata, and unfortunately for Tobey his penis is severed. At this point in the film, understandably traumatized, Dawn believes Tobey's death is her fault. Herein lies yet another fallacy exposed by *Teeth*. Even though Tobey entered Dawn's body without her permission, Dawn blames herself. This is rape culture. This is the toxic sensibility that a woman's destruction rests on the length of her skirt. It is unfortunate that certain horror movies have perpetuated this way of thinking, though movies like *Teeth* and *It Follows* (2014) question these female punishments.

In *It Follows,* Jay (Maika Monroe) has consensual sex with her boyfriend, Hugh (Jake Weary). Immediately after, she is chloroformed and tied up. This juxtaposition of sex and violence is not an example of the trope, but an observation on what we have become used to in horror. Jay soon finds out that she has been given something far worse than a traditional sexually transmitted infection. Her punishment is that of a supernatural entity, one that will follow her, and eventually kill her, if she doesn't pass it on. Jay has been given the onus of sex by a man. He is able to disappear into the night, leaving her with the product of their shared union. This could allegorically be tied to everything from the aforementioned STIs, to pregnancy, to the shame, guilt, and ownership of not remaining "pure." In his article for *Bloody Disgusting*, Brendan Morrow points to another symbol within the film's context, that it is really about sexual assault: "What's happening to Jay is less an analogue to having an STD and more a metaphor for life as a rape survivor. Even though the sex was consensual, the image of Hugh knocking Jay out certainly calls rape to mind, and besides, could Jay *really* give consent without having any knowledge of the creature?"[8]

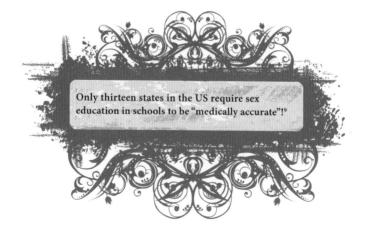

Only thirteen states in the US require sex education in schools to be "medically accurate"![9]

As *Teeth* progresses, Dawn learns more about her sexuality. She removes the sticker from the textbook, studying the image. She has consensual sex, understanding that it is the consent that stops the dentata from slicing. And, in the end, she uses her sex as a weapon against the men who deserve it. This reversal, from Dawn's fear of her own desire to a woman with complete agency over her body, makes *Teeth* a must-watch feminist film.

CHAPTER NINE
GERALD'S GAME

Year of Release: 2017	
Director: Mike Flanagan	
Writer: Jeff Howard, Mike Flanagan	
Starring: Carla Gugino, Bruce Greenwood	
Network: Netflix	

Hormonal teenagers run rampant in horror media, finding all sorts of dangerous places for a quickie. *Gerald's Game* (2017), based on the 1992 novel by Stephen King, isn't a story about them. It's about womanhood. It's about marriage. It's about a woman facing the knowledge that she has been tethered her entire life.

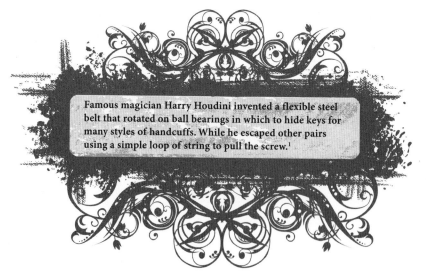

Famous magician Harry Houdini invented a flexible steel belt that rotated on ball bearings in which to hide keys for many styles of handcuffs. While he escaped other pairs using a simple loop of string to pull the screw.[1]

Jess (Carla Gugino) waits on a bed for her husband, Gerald (Bruce Greenwood). She yanks off the tag of her new lingerie, primping her hair. We've already seen, subtly by director Mike Flanagan, that despite her appearance, she is not excited about the sexual encounter about to occur. This becomes more and more clear as Gerald handcuffs her to the bed, enacting a rape fantasy in which Jess is not a willing partner. When Gerald suddenly falls dead from a heart attack, we see that Jess is helplessly tied to the bed, but at least she is alone. Unmolested.

Sexual assault is a common plot device in all genres, unfortunately because it is a sobering reality, as in the US one in three women and one in six men have experienced some form of sexual violence in their lifetime.[2] In *Gerald's Game*, we witness what begins as a consensual encounter between Jess and Gerald turn into an unwanted sexual attack. This marital rape between

The left hand features a person who has acromegaly compared to someone who does not. Carel Struycken, who portrays Moonlight Man in *Gerald's Game*, has the disease. This condition, caused by excess growth hormone, is the reason for the actor's 7-foot height and distinctive facial features.

spouses, not so long ago considered legal, is the kind of unsettling and stark reality that is not a popular topic in horror. One reason for this is perhaps that women's sexuality in media versus teens' is far less prevalent. As Jess's memories are revealed, we find out that she was also assaulted by her father at age twelve and then, worse in her estimation, manipulated to keep it a secret.

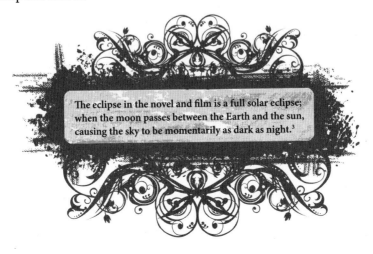

The eclipse in the novel and film is a full solar eclipse; when the moon passes between the Earth and the sun, causing the sky to be momentarily as dark as night.[3]

When Jess speaks to a physical manifestation of her dead husband, they recount a "joke" Jess overheard her husband tell. When asked "What is a woman?" Gerald smugly responds, "A life support system for a cunt." She did nothing at the time, as, per her father, she'd learned to suppress. Her self, her sexuality were not her own.

"Wake up." Jess pleads with herself. "You've been sleepwalking since you were twelve."

While *Gerald's Game* is strikingly feminist in its treatment of women's sexuality, there are other horror movies that glorify the brutalization of naked, sexualized females. In her piece for *American Horrors: Essays on the Modern Horror Film*, Robin Wood gives insight into why certain horror films with a male perspective revel in sexualized violence: "The violence against women in movies have generally been explained as a hysterical response to 1960s and 1970s feminism: the male spectator enjoys a sadistic revenge on women who refuse to slot neatly and obligingly into his patriarchally predetermined view of the way things should naturally be."[4]

Women have responded to this by creating their own films, taking back the power of the sex assault narrative. In *American Mary* (2012), writers and directors Jen and Sylvia Soska highlight the gravity of sexual assault to a woman's psyche.

Female beauty and the vanity it supposedly inspires have been a part of fiction since before Snow White was put in a coma for being too pretty. But, to understand its sexualized place in the slasher genre, one must accept that many of the traditional films are shot with a male gaze. In "Masculinity and Monstrosity: Characterization and Identification in the Slasher Film," Klaus Rieser describes fellow researcher Laura Mulvey's piece *Visual Pleasure and Narrative Cinema*:

> In it, she uncovered an ideological imbalance between male and female, which in the classic film text are constructed visually as "gazing/to-be-looked-at-ness" and narratively as "active/passive." The spectator implied (or indeed constructed) by this process is male, and his visual pleasures are divided between a fetishistic or sadistic/voyeuristic gaze on the female and a narcissistic

identification with the male characters. Through techniques such as invisible editing and subjective camera, an essentially sadistic position vis-à-vis the story's female character (either "rescuing" or punishing the woman for her desire) is carved out for the male spectator.[5]

In other words, these female characters' attractiveness is fetishized for the male audience, not for their own sexual control or pleasure. Just as Jess wore her lingerie not because she wanted to, but because she felt she had to.

Woman's control over her sexuality, as described in "Open Body/ Closed Space: The Transformation of Female Sexuality," is a concept that has been handled quite differently from men's: "The key to attitudes regarding men and women is the belief that the sexual drive in the adult female is subject to her control, while that of the adult male is physiologically imperative and cannot be controlled."[6] This unfounded belief, that men can't help themselves, no doubt contributes to rape culture. *Gerald's Game* explores these long-held notions, asking who is in real control of Jess's sexuality? And who should be?

Before finding her strength and escaping both her literal and metaphorical tethering, Jess bemoans her tragedy: "Gerald's five inches, [my] life has to add up to more than that!" In the end, Jess makes certain that happens.

SECTION FOUR
REVENGE

CHAPTER TEN

US

Year of Release: 2019	
Director: Jordan Peele	
Writer: Jordan Peele	
Starring: Lupita Nyong'o, Winston Duke	
Budget: $20 million	
Box Office: $249.6 million	

Revenge is a concept that has been around forever. The Bible, in Exodus 21:23, instructs, "you are to take life for life, eye for eye, tooth for tooth, hand for hand, foot for foot, burn for burn, wound for wound, bruise for bruise." William Shakespeare wrote of revenge in *Hamlet* (1599). Commanded by his father's ghost in Act One to "revenge his foul and most unnatural murder" by his brother Claudius, who has robbed him of his wife and throne as well as his life, Hamlet swears that "with wings as swift / As meditation, or the thoughts of love," he will "sweep to [his] revenge."

Revenge is also a theme in art. One of the most famous paintings featuring revenge is *Prometheus Bound* (1612) by Peter Paul Rubens, in which Prometheus receives punishment after giving the gift of fire to mankind. This is considered a lesson from mythology: "there's no greater treason than the one committed toward a deity, particularly a Greek god, since they're known to be the most vengeful beings of history."[1] Another painting, *Judith and Holofernes* (1599) by Caravaggio, shows us what a person is capable of doing in order to save their people. It depicts, in brutal and bloody detail, Judith decapitating Holofernes while he is drunk. She does this in order to avoid the invasion of her hometown, Betulia, from total destruction.

The long history of vengeance in art and literature suggests a basic human instinct for retribution that is engrained in the mind or spirit.

Facts confirm this common need for revenge as it has been cited as a factor in one in five murders that occur in developed countries. A report from 2002 found that between 1974 and 2000, three in five school shootings in the United States were driven by vengeance.[2] This is not a new concept but one that seems to be eerily present throughout history.

Social psychologist Ian McKee, PhD, of Adelaide University in Australia studies what makes a person seek revenge rather than just letting an issue go. In June of 2008, he published a paper in *Social Justice Research* linking vengeful tendencies primarily with two social attitudes, right-wing authoritarianism and social dominance, and the motivational values that underlie those attitudes: "People who are more vengeful tend to be those who are motivated by power, by authority and by the desire for status. They don't want to lose face."[3] The study suggests that those who value power will feel like their power is affected when a crime or infraction is committed against them. These people are more likely to seek revenge or get even in some way for the perceived damage that was done to them. Those who are more focused on tolerance, understanding, and the welfare of others are less likely to dwell on the thought of revenge or seek compensation for damages in the same incidents or encounters.

Those who follow through with avenging situations when they feel slighted aren't necessarily happier in the long run. If revenge doesn't make us feel any better, why do we seek it? One hypothesis states, "punishing others . . . is a way to keep societies working smoothly. You're willing to sacrifice your well-being in order to punish someone who misbehaved."[4] Societies and cultures throughout the world have taken it upon themselves to take revenge on those they feel have done wrong. Whether for good or for bad, it's a very real concept with some potentially scary consequences.

When delving into the subgenre of revenge in horror, you may discover that many films tackle the subject of rape revenge. Rape is often used as a backstory, or tool of empowerment for women in storytelling, but film scholars question if it has to be. Jill Gutowitz said, "Many writers do their female protagonists justice, and rape-revenge stories can be mollifying to watch, especially for survivors of assault. But film and TV characters are a reflection of real-life women, and there are an infinite amount of motivations, methodologies, and histories that produce interesting, complicated women."[5]

We want to explore the concept of revenge in horror films from a different perspective. What are those other, unique backstories of women seeking revenge? How do they differ from their male counterparts? One of the most original and thrilling tales of revenge is the film *Us* (2019). Written and directed by Jordan Peele, *Us* follows the story of two women, Adelaide and Red, both played by Lupita Nyong'o. As we discover in the end of the film, the girls switch places one night at an amusement park. This moment in the plot touches on an idea in folklore called a changeling, a fairy child left in place of a human child stolen by the fairies. The theme of the swapped child is common in medieval literature and reflects

Scissors were thought to ward off evil beings.

concern over infants thought to be afflicted with unexplained diseases, disorders, or developmental disabilities. Simple charms were thought to ward them off such as an inverted coat or open iron scissors left where the child sleeps. The doppelgängers in *Us* use gold scissors as weapons throughout the film.

Do doppelgängers exist in reality? A doppelgänger is a non-biologically related lookalike or double of a living person, sometimes portrayed as a ghostly or paranormal phenomenon and usually seen as a harbinger of bad luck. Other traditions and stories equate a doppelgänger with an evil twin. The term is used to describe any person who physically resembles another person. The concept of alter egos and double spirits has appeared in the folklore, myths, religious concepts, and traditions of many cultures throughout human history. In Ancient Egyptian mythology, a "ka" was a tangible spirit double having the same memories and feelings as the person to whom the counterpart belongs.

Heautoscopy is a term used in psychiatry and neurology for the reduplicative hallucination of "seeing one's own body at a distance."[6] It can occur as a symptom in schizophrenia and epilepsy. Heautoscopy is considered a possible explanation for doppelgänger phenomena and can often be seen as a harbinger of death. One case studied in 1884 followed a man who, at first, saw his doppelgänger as benevolent but eventually the double became more autonomous and the man shot himself after being humiliated by his "other." Another case in 1994 studied a man fighting his own doppelgänger. It perhaps looked something like the scene in *Evil Dead 2* (1987) when Ash (Bruce Campbell) is fighting his own possessed hand. The man jumped out of a third-story window in order to stop the feeling of being divided in two.[7]

Nature versus nurture is another topic that can be explored in relation to the film *Us*. A documentary called *Three Identical Strangers* (2018) examined the story of triplets who were separated at birth and each adopted into separate families with different backgrounds and in varying socioeconomic classes. Researchers followed their progress as they aged to see if nature or nurture affected their lives, choices, health, and personalities. The findings of the study were inconclusive, and the study itself, it was revealed, was never approved. This ties into the concepts of privilege and opportunity in the film *Us* and how they separate the tethered from the untethered. Those who live above ground in the film are able to make choices, have agency, and live their lives according to how they see fit. The tethered, who live underground, aren't allowed the same opportunities. They are forced to follow the path laid out for them, have no agency, and must blindly follow where they are led. The escalator to the underground only goes down, not up. This can be seen as a metaphor for privilege and how difficult it can be in our society to get past the gatekeepers or hurdles that are placed in the way.

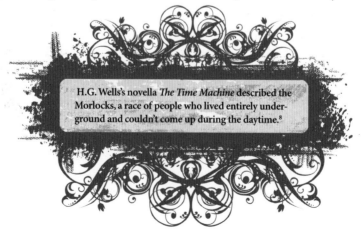

H.G. Wells's novella *The Time Machine* described the Morlocks, a race of people who lived entirely underground and couldn't come up during the daytime.[8]

Class is defined as "the status an individual or group achieves by virtue of its economic strength, the influence among other groups, and the power to affect change in its community of choice."[9] Classism is the "systematic oppression of subordinated groups (people without endowed or acquired economic power, social influence, and privilege)" and is held in place by a system of beliefs that "ranks people according to economic status, family lineage, job, and level of education." Class is represented by the haves and the have-nots in *Us*: "Classism says that dominant group members are smarter and more articulate than working class subordinated groups. In this way, dominant group members (upper middle class and wealthy people) define for everyone else what is 'normal' or 'acceptable' in the class hierarchy."[10]

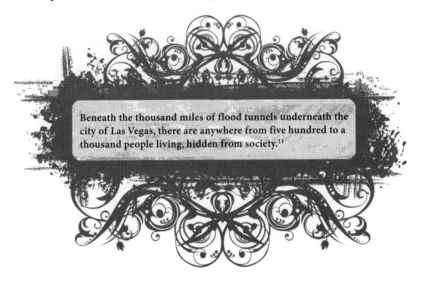

Beneath the thousand miles of flood tunnels underneath the city of Las Vegas, there are anywhere from five hundred to a thousand people living, hidden from society.[11]

The revenge plot in *Us* is based on the character Red's prior knowledge of the world above; *C.H.U.D.* (1984), *The Goonies* (1985), and Hands Across America (1986) are hints we are given from Red's childhood and the media she was exposed to. The two movies, seen as VHS tapes in the first scene of the film, both explore this dynamic of class and privilege and focus on underground elements. These themes and social commentary are a welcome and exciting addition to the movie *Us*. How Red decides to handle her anger and frustration from decades of living in unfair conditions is central to her character's journey. Although the idea of seeking revenge may seem cathartic, experts argue that there are more useful ways to handle these complicated emotions. Aristotle used the term *katharsis* to describe the therapeutic effects of theatrical tragedy in purging emotions, while Sigmund Freud described it as the process of reexperiencing repressed emotions in therapy.

Studies show that those who seek revenge often have higher levels of aggression than those who do nothing at all. According to a study by Brad J. Bushman, "Venting to reduce anger is like using gasoline to put out a fire—it only feeds the flame. By fueling aggressive thoughts and feelings, venting also increases aggressive responding."[12] Red may have had a lot of time on her hands to plot her revenge against the world above, but we shouldn't be encouraged to do the same. Instead, we should seek out our "forgiveness instinct" in order to "suppress the desire for revenge and signal our willingness to continue on."[13] Did the tethered feel better after ascending from the underground and killing their contemporaries? Perhaps. And that's what makes *Us* a terrifying horror film.

CHAPTER ELEVEN

WHAT LIES BENEATH

Year of Release: 2000	
Director: Robert Zemeckis	
Writers: Sarah Kernochan, Clark Gregg	
Starring: Michelle Pfeiffer, Harrison Ford	
Budget: $100 million	
Box Office: $291.4 million	

Vengeful ghosts have haunted readers and campfire storytellers for centuries. It is said in folklore that the dead return to avenge their wrongful or unjust deaths, haunt those who harmed them in life, or seek revenge for improper burials. Most often in these tales, the vengeful ghost is a woman who was wronged in life and will walk the Earth until she feels retribution. "The Green Lady" is one such woman and is believed to haunt castles in Scotland. In 1920, the skeleton of a woman was found in Fyvie castle. After she was buried, the residents of the castle reported hearing strange noises and experienced unusual occurrences. The body was exhumed and placed back in the castle, and the strange happenings, coincidentally, ceased.

Today, there are an estimated 1,200 haunted house attractions in the US, and an average of 8,000 people attend them every year.[1]

The "Chindi" of Navajo folklore are thought to be the ghosts left behind after people die. They contain all of the bad things about the individual from their past. This vengeful ghost is believed to be the cause of "ghost sickness" if anyone comes into contact with it. The symptoms of ghost sickness include feeling weak, having a loss of appetite, experiencing the feeling of being suffocated, and having recurring nightmares. This spirit can also be seen in dust devils that swirl counterclockwise in nature.

The Curse of La Llorona (2019) is based on the myth in Mexico of "The Woman in White" or "The Weeping Woman." She is a mother who drowned her own children after discovering that her husband was cheating on her. Consumed by grief, her ghost can be heard crying as she searches for other living children to replace her own. The legend is popular in Mexico and some areas of the southwestern United States. People who believe in La Llorona keep her away by lighting candles, hanging crosses, and saying their prayers.

"The Chudail" of India is another myth of a female vengeful ghost. According to some legends, she is a woman who died during childbirth or pregnancy or due to suffering at the hands of her in-laws and will come back to seek revenge, particularly targeting the males in her family. She is described as hideous but "has the power to shape-shift and disguise herself as a beautiful woman to lure men to the mountains where she either kills them or sucks up their virility, turning them into old men."[2]

The Ring (2004) is based on the Japanese folktale of a vengeful ghost and retells the story through a modern lens. The character of Samara comes back from the grave to haunt and kill anyone who watches the videotape that ultimately hints to her brutal demise. She gets her revenge not only on those who harmed her in life, but on those she deems worthy of it: those who don't empathize with her situation or share her burden.

The ghost in *The Ring* is based on Japanese folklore.

The *Ju-On* film series (2003–present) follows the stories of those affected by a curse and the ghost Kayako who marks and pursues anyone who enters the house where she dwells. *Shutter* (2004) begins with a couple who flee from a hit-and-run accident. The girl who died as a result comes back to haunt the couple, appearing in photographs and manifesting herself as a physical pain on the man's neck.

The haunted attraction industry makes between $500 and $800 million a year in ticket sales.[3]

This story trope of a woman haunting others from beyond the grave has also been seen in literature such as Edgar Allan Poe's *The Pit and the Pendulum* (1842), *Rebecca* (1938) by Daphne du Maurier, and *Beloved* (1987) by Toni Morrison. As author Tabitha King said, "The ghost is almost always a metaphor for the past." This metaphor is exactly what is haunting Norman (Harrison Ford) and his wife in *What Lies Beneath* (2000). Claire (Michelle Pfeiffer) suffers from empty nest syndrome when her daughter leaves for college. She is now alone, in a big house, a year after suffering injuries from a car accident. We are given hints about the accident and Claire's backstory throughout the film, and it's an eerie feeling to know that her memories have to be filled in by those we're not sure we can trust. Although empty nest syndrome isn't considered a clinical condition, its symptoms are real and include depression, sense of loss or purpose, stress, and anxiety.

Gaslighting is a term used to describe a type of abuse suffered by those who are being told what to believe or whose feelings are dismissed and invalidated. Claire, by not remembering the details of her husband's affair and her accident, is being gaslighted by Norman. She's made to feel silly for thinking there could be anything going on in her house or with her neighbors next door. Claire is worried that Mary (Miranda Otto) is in an abusive relationship, and she even fears that Mary's husband, Warren (James Remar), has murdered, or will murder, his wife. According to Sigmund Freud, projecting onto others is a psychological defense mechanism that we use to cope with undesirable emotions. Claire, through her foil of the neighbor, saw herself reflected back. Mary was also married to a successful professor, spending most of her day alone in a big, empty house. Apart from projection, other common coping mechanisms include denial (refusing to admit to yourself that something is real), distortion (changing the reality of a situation to suit your needs), passive aggression (indirectly acting out your aggression in a subtle way), repression (covering up feelings or emotions instead of coming to terms with them), sublimation (converting negative feelings into positive actions), and dissociation (substantially but temporarily changing your personality to avoid feeling emotion). Claire and Norman seem to carry out each of these coping mechanisms throughout the film.

Giving up dreams, or a career, for marriage or motherhood is something that women often do in real life and is explored through film and literature. Claire, in *What Lies Beneath*, was a successful musician who gave up touring and fame to be a wife and mother. The Pew Research Center reports that 10 percent of highly educated mothers (those who earned a master's degree or greater) stay home with children instead of continuing their careers. Some may choose to do this for financial reasons, while others do it for altruistic ones. Regardless of the reason, women often struggle with this decision and may feel guilt over the path they choose. *What Lies Beneath*'s cowriter Sarah Kernochan said:

I think people know that there's no point in calling me in if you want the other kind of women characters: a featureless "help me" character, or the saint, the whore—you know, any of the archetypes. I don't think all women are powerful, intelligent, any of those things. I just require that female characters be very real, that they have all the dimensions that the male characters do.[4]

Claire is absolutely one of those characters in the film and is able to save herself multiple times throughout the story.

Why do we love vengeful ghost stories so much? Maybe it's because we like to be scared by the unknown. Maybe the thought of ghosts gives us hope for an afterlife. Or maybe we secretly think that if someone wronged us in life, we could come back to haunt them. Whatever the reason, ghosts out to get their revenge are sure to be here to scare us for years to come.

CHAPTER TWELVE

A GIRL WALKS HOME ALONE AT NIGHT

Year of Release: 2014	
Director: Ana Lily Amirpour	
Writer: Ana Lily Amirpour	
Starring: Sheila Vand, Arash Marandi	
Budget: $56,000	
Box Office: $628,000	

In 1871, the novella *Carmilla* by Joseph Sheridan Le Fanu was released and told the story of a young woman being stalked by a female vampire. It was a story centered on women with sexual and seductive undertones, perhaps too risqué for the time. Although it was considered a flop, it did go on to influence Bram Stoker in writing *Dracula* (1897). As U. Melissa Anyiwo notes in her essay "The Female Vampire in Popular Culture," *Carmilla* "represents the most complete vision of the female vampire: a lesbian seductress who desires to overturn patriarchy by promoting female independence from men, and the rejection of biological reproduction."[1] Vampires in horror movies are often shown as sexual creatures. *Bram Stoker's Dracula* (1992) features several scenes of female vampires, known as "Brides of Dracula," luring men into their traps. These characters may use their feminine wiles to lure and prey on others. Xan Cassavetes, who directed *Kiss of the Damned* (2012), said:

> I think of women usually being the alluring and powerful preda-
> tors. I am as enchanted by a beautiful woman as anyone and see
> them as mysterious creatures who often don't know their own

power or how to navigate it which is so interesting and touching to me. A woman who is either distinctly moral or immoral is always fascinating to me and this film has characters who are both, although this does not mean one is the outright villain and one isn't. There is a fine line between integrity and righteousness just as there is between realism and cynicism and I liked exploring this through female characters, who never really are portrayed as having these dilemmas or qualities as much as male ones.[2]

A Girl Walks Home Alone at Night (2014) explores a female vampire character who is unlike the overly sexualized vampires of other films. The movie is described as the first Iranian vampire Western. The title alone implies the Girl, played by Sheila Vand, may be weak or in danger. But the film subverts that: the Girl is dangerous. *She* is empowered. She is not the prey but the predator. Writer and director Ana Lily Amirpour said, "A vampire is a serial killer, a historian, an addict, and a romantic, all in one. What filmmaker wouldn't want to explore this mythical creature?"[3]

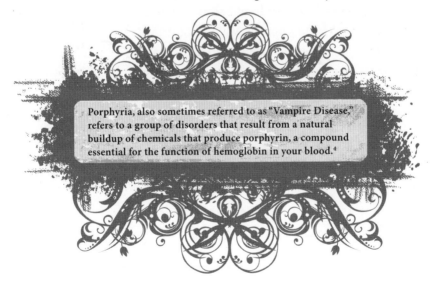

Porphyria, also sometimes referred to as "Vampire Disease," refers to a group of disorders that result from a natural buildup of chemicals that produce porphyrin, a compound essential for the function of hemoglobin in your blood.[4]

The stereotypical patriarchy is shown through several scenes as the film begins: a man on the television is informing women of their duty to cook and clean for their husbands as they work and make money for them. A drug dealer/pimp character assumes one of his workers must want children soon, since she's thirty and that's "getting old," even calling her a "hag." The first time we see the Girl, she is dressed in a chador, an open cloak worn by many women in Iran. Viewed by some as oppressive, it becomes a symbol of her strength. Amirpour said, "In Iran, I have had to wear a hijab [headscarf], and personally I find it completely suffocating. I don't want to be covered up in all that cloth. But there was something about the chador, though. It's made of a different fabric. It's soft and silky and it catches the air. When I put it on, I felt supernatural. But I also get to take it off."[5]

A poster of Margaret Atwood as Madonna adorns the wall in the Girl's bedroom, and we get a glimpse into her life before we discover that she is a vampire out for revenge. The pimp does drugs in front of her, counts money, and lifts weights in a house decorated with taxidermied animals: all signs of traditional male dominance and intimidation. But the Girl doesn't flinch: "In her own way, the Girl is pursuing a campaign of revenge against the theocratic oppression of women, and also against the more familiar Westernized version of misogyny . . ."[6] She kills the man and drains him of his blood before leaving.

A Girl Walks Home Alone at Night is described as part Western, and the look and feel of the film supports that description. There are a handful of horror movies that fit into the Western genre including *Sundown: The Vampire in Retreat* (1989), *Vampires* (1998), and *The Burrowers* (2008), but more traditional Western movies often have a theme of revenge running through their plot. *Unforgiven* (1992) follows the story of a gun-for-hire getting revenge on a pair of cowboys for attacking a woman. *The Quick and the Dead* (1995) focuses on a female gunslinger coming to a town to get her revenge on those who killed her father. How does *A Girl Walks Home Alone at*

The duel is a classic trope in Westerns.

Night fit into the traditional Western tropes? Westerns are usually set within a society organized around codes of honor and direct justice being used to uphold those codes. Social order is maintained by a cowboy or gunfighter, often depicting a showdown or duel in the street at high noon. Several shots in *A Girl Walks Home Alone at Night* harken to this visual of a standoff. As the Girl stalks her prey, she stands stoically some distance away from them. Like a gunfighter waiting to draw his weapon from the holster, the Girl waits to attack. Her body (or mouth specifically) is her weapon, and the attack becomes more personal.

Like many classic Westerns, the film is set in a small, desolate town where crime is running rampant. Sergio Leone, one of the legends of the Western filmmaking genre, described these towns as places where life has no value. Life does have value to the Girl, but perhaps not in the same way as with a traditional Western hero. She kills those who take advantage of, or harm, others and uses them for sustenance. She is like many Western heroes; she has questionable morals and a mysterious origin. She, like her male counterparts in other films, doesn't necessarily show emotion, especially in front of others.

The Cowboy Syndrome is a phenomenon in which people feel like they can't express their true feelings or show emotion due to society's judgment of them. The name comes from those traditional John Wayne movies: cowboys don't cry. They're tough, macho, and focus more on their feelings of anger or revenge. Studies show that this isn't a healthy way to live.[7] The syndrome is most predominant in the United States with males but has shifted to include women feeling like they can't express their emotions in public. The Girl in *A Girl Walks Home Alone at Night* doesn't necessarily have the Cowboy Syndrome, but she does seem to repress her emotional expression.

In traditional movies in the Western genre, women are portrayed as damsels in distress. The Girl in this film needs no one to save her. The film is not only horror and Western, but also contains a love story. The Girl meets Arash (Arash Marandi) one night on the street while he is dressed as Dracula. It's a meet-cute for the ages as he assumes she may be afraid of him: a common but real fear. Instead, she looks out for him even though she is the true threat. Meg and I (Kelly) both love a tale of forbidden romance! The Girl and Arash are both flawed but are trying to follow their own moral codes as best they can. As they drive off to begin their new lives at the end of the film, we are left to wonder: can they make it work? "The Girl is both the final girl and the villain of her own horror movie, a romantic interest to the eyes of some and something to be afraid of to the eyes of others: a complex female character that is in need of no man to save her."[8]

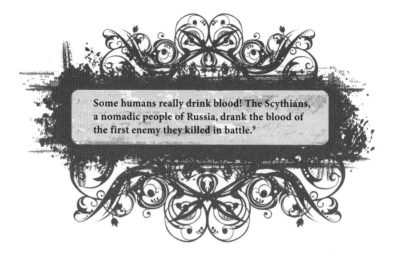

Some humans really drink blood! The Scythians, a nomadic people of Russia, drank the blood of the first enemy they killed in battle.[9]

Many movies in the horror genre feature humans getting revenge on vampires for killing their loved ones. In *Abraham Lincoln Vampire Hunter* (2012), the title character vows to fight the undead after losing his mother to a vampire's bite. *Blade* (1998), although half vampire himself, is also out to avenge his mother's death. It's rare that a vampire seeks revenge on humans for their misdeeds, but *A Girl Walks Home Alone at Night* does just that. Revenge as a theme will undoubtedly continue to permeate art and media for centuries to come. As Machiavelli said, "People should either be caressed or crushed. If you do them minor damage, they will get their revenge, but if you cripple them, there is nothing they can do. If you need to injure someone, do it in such a way that you do not have to fear their vengeance."[10] Numerous horror movies have been based on this premise and explore this human instinct. We, as the audience, may derive some pleasure from it. Stephen King wrote, "For people like us, little people who went scurrying through the world like mice in a cartoon, sometimes laughing at the assholes was the only revenge you could ever get."[11]

SECTION FIVE
THE INNOCENT

CHAPTER THIRTEEN
LET THE RIGHT ONE IN

Year of Release: 2008	
Director: Tomas Alfredson	
Writer: John Ajvide Lindqvist	
Starring: Kare Hedebrant, Lina Leandersson	
Budget: $4.5 million	
Box Office: $11.2 million	

They say that little girls are made of sugar and spice and everything nice. Normally, that is very true, but horror is a place of what-ifs, of exploring the cracks in the perfect veneer. And what is more perfect than a young and beautiful little girl? This seeming perfection comes from our cultural belief that children are blank slates, unable to interpret or become a part of concepts like adult cynicism or evil.

As we'll explore, this female innocence trope begins with little girls and extends to adolescence and even to adulthood, a length of time not normally attributed to males. Before we discuss the subversion of innocence, we must first point to examples of innocent girls who remain how we typically picture them. A prime example would be sweet, little Carol Ann (Heather O'Rourke) in *Poltergeist* (1982). Carol Ann is at the pinnacle of innocence throughout the movie. At five, she is the youngest child in the Freeling family and is very much concerned with the thoughts of the young. She sleeps with a stuffed animal, speaks softly, and can't fathom why her parents are so upset about the recent disturbances in their new home. She is made of pure, youthful energy. Tangina (Zelda Rubinstein) explains how this innocence affects the dead souls she is with across the celestial plane: "They are attracted to the one thing about her that is different from themselves: her life force. It is very strong. It gives off its own illumination. It is a light that implies life and memory of love and home and earthly pleasures." Even after her abduction to the "other side," Carol Ann maintains her innocence. The dead are not able to infect her with their darkness.

Other innocent young girls are not as lucky. Regan (Linda Blair), of *The Exorcist* (1973), is only twelve years old when her body is ravaged by an invading demon from hell. Regan does not choose this path and is not in control of what she does. Afterward, we see the physical toll this possession has taken. The use of Regan, particularly her portrayal as an innocent girl before the transition, is a well-calculated plot device to further the audience's horror. Would *The Exorcist* have had the

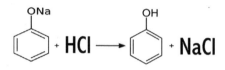

Hakan (Per Ragnar) uses hydrochloric acid to disfigure his own face. Used mainly to refine metals, hydrochloric acid is extremely toxic. Even its mist can corrode skin.[1]

same effect if a grown-up, mired in the complexities of adulthood, had used such vile language toward a priest? This juxtaposition of a pure girl in the hands of the devil has been further used to great effect in films like *Ouija: Origin of Evil* (2016) and *The Conjuring Two* (2016). Girls like Regan and Carol Ann are victims of circumstance, perhaps even chosen because of their trusting, naive natures. But what about the girls who are innately evil? The ones who have agency over their choice to maim, kill, and wreak havoc? This subversion is the most popular element of the trope, populating numerous movies, television series, and books. In *The Uncanny Factor: Why Little Girls Scare the Shit Out of Us*, Robbie Blair proposes the theory that because of our natural inclination to see little girls as helpless, girls who are not so simply scare us:

> It's honestly hard to say whether it's our evolutionary heritage or our cultural standards that leads us to feel so protective and nurturing toward little girls, but the desire to take care of them certainly seems to be the immediate instinct. Even when the young girl is not familiar to us, we have a sense of familiarity with what she represents—and the vulnerability of the little girl makes us more willing to be vulnerable ourselves.[2]

This vulnerability leads us to feel the ultimate chill when little girls do not meet our expectations. In her article "The Terror of Little Girls: Social Anxiety About Women in Horrifying Girlhood," Leigh Kolb counters that why we are fascinated with these "demonic" little girls is rooted in the female:

> We wouldn't be so shaken to the core by possessed, haunted, violent little girls if we were simply supposed to be longing for innocent times of yesteryear. Instead, these little girls embody society's growing fears of female power and independence. Fearing a young girl is the antithesis of what we are taught—stories of missing, kidnapped, or sexually abused girls (at least white girls) get far more news coverage and mass sympathy than stories of boy victims. Little girls are innocent victims and need protection.[3]

Kolb further asserts that after a long period in horror cinema of woman as victim, a comforting role for audiences to comprehend based on our societal expectations, the advent of women's rights brought about the barrage of scary and unpredictable little girls: "It's no coincidence that the fifties and sixties were seeing sweeping social change in America (the Pill, changing divorce laws, resurgence of the ERA, a lead-up to Roe v. Wade)." Thus, *The Bad Seed* (1956) became one of the first cinematic depictions of this innocent facade of a small girl crumbling away to reveal innate evil. Based on the 1954 novel and subsequent play, *The Bad Seed*, directed by Mervyn LeRoy, centers on eight-year-old Rhoda Penmark (Patty McCormack). On the outside, Rhoda embodies everything a mid-twentieth-century girl ought to. She has blonde, well-kept braids, a sunny smile, and can even tap dance. In one scene, Leroy (Henry Jones) confronts Rhoda about a boy she hit with a stick. As Leroy recounts the violent act that he knows Rhoda is responsible for, the little girl, dressed in frills, plays with a tea set. He explains that he knows she's mean, because he is mean himself. This reflection, of man to girl, is vital to note, as it may provide insight into why adults find "creepy" girls so universally terrifying. Because, you see, Rhoda gets her revenge on Leroy, burning him to death. Thus, she has done what the women in her orbit, like her homemaker mother, struggle to do: show dominance by usurping the male. Rhoda has proven *she* is meaner.

Another girl who appears sweet as pie is Esther (Isabelle Fuhrman) in *Orphan* (2009). Like Rhoda, Esther, who is adopted as an older child, displays the outward femininity that society expects. Yet, she also lashes out violently, resorting to murder. The film's shocking reveal, that Esther is not in fact a little girl, but rather a woman who suffers from a disease that makes her *look* like a girl, is vital in understanding the pressure we put on girls to exist within what we believe to be a space of purity and femininity. As noted in "Gendered Social Control: 'a Virtuous Girl' and 'a Proper Boy,'" when girls or women deviate from what we expect, however subtle or extreme, we as a society try to usher them back to what is deemed normal female behavior. The social control of women considered deviant has frequently been described, by feminist crimi-nologists, as deeply gendered: the social control of women and girls is aimed at making them conform to traditional femininity.[4]

On the far end of the spectrum, a girl who seems to be born evil and remains so after her death is Samara (Daveigh Chase) from *The Ring* (2002) or Sadako (Rie Ino) from the Japanese original, *Ringu* (1998). When Rachel (Naomi Watts) discovers that Samara's mother threw her down a well, leaving her to starve to death, she has empathy for the little girl. She even excuses the ghost's violent impulses, holding her skeletal remains maternally as a sign of closure and goodwill. But Rachel and the audience soon find out that Samara is not the poor, abused girl she is believed to be. Her thirst for death does not stop. This is when the true terror settles in, that Samara cannot be easily categorized as a trauma-tized little girl, lashing out. She chooses her behaviors and is willfully destructive and cruel. And, like her Japanese counterpart, Samara seeks revenge under a curtain of unkempt hair. Girls like Esther and Rhoda weren't so bold as to rebel in their looks. Perhaps because Samara is dead, she can fully embrace her rejection of girlhood beauty standards.

In the 2008 Swedish film *Let the Right One In*, the character of Eli (Lina Leandersson) is not quite as simple as Carol Ann, or as evil as Samara. It is this complexity that provides an empathetic and authentic girl. A girl who just happens to be a vampire.

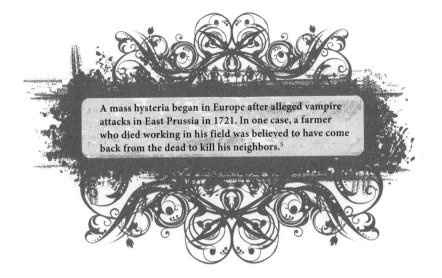

A mass hysteria began in Europe after alleged vampire attacks in East Prussia in 1721. In one case, a farmer who died working in his field was believed to have come back from the dead to kill his neighbors.[5]

The film, directed by Tomas Alfredson, opened to immediate praise from critics, winning multiple awards ranging from Best Narrative Feature at the Tribeca Film Festival to Best Horror Film at the Empire Awards. It was so wildly popular at the time that an American remake was soon developed, *Let Me In* (2010), starring Chloe Grace Moretz.

The main character in *Let the Right One In* is Oskar (Kare Hedebrant), a timid, bullied twelve-year-old boy who finds a friend in Eli. Eli appears to be normal, perhaps even someone to be pitied, as she does not have the glossy appearance of *The Bad Seed*'s Rhoda. Like any child would, she finds interest in Oskar's Rubik's Cube, though she demonstrates superior intelligence in quickly figuring it out. She also seems to have a genuine affection for Oskar, which, in stark contrast to the vicious attacks she carries out, gives Eli a soft, innocent side.

As their relationship develops, they have a sweet moment, lying side by side, in which Oskar asks Eli if she'll "go steady with him." When Eli asks what that would be like, Oskar assures her it is a normal thing for boys and girls to do together. To which Eli simply states, "I'm not a girl."

Because of her vampirism, Eli's gender and age are up for speculation, but what is important to note is that once again there is the concept of a girl not being a girl, and how this horrifies. Like Esther, Eli admits that she is not what she appears to be. This harks back to the gender fluidity

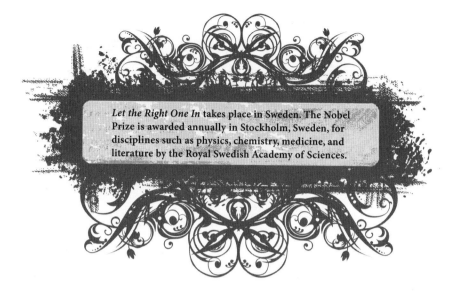

Let the Right One In takes place in Sweden. The Nobel Prize is awarded annually in Stockholm, Sweden, for disciplines such as physics, chemistry, medicine, and literature by the Royal Swedish Academy of Sciences.

of Angela (Felissa Rose) in 1983's *Sleepaway Camp* (discussed in Chapter Seven on p. 47) and how, once again, these disparate aspects cause fear. Whether a girl is inhabited by a demon, really an adult, or controlled by her craving for blood, it is the "uncanny" or the subversion of our expectations that we as a society struggle to comprehend. Eli could be the poster girl for the creepy little girl trope, because she walks an ambiguous line. Unlike Rhoda or Samara, she is not pure evil. She is both a victim of her circumstance (the need to subsist on blood) as well as a female (or not?) with agency, leading Oskar down a road of violence in order to assist her in what she needs. Once we feel a moment of empathy for Eli, we are again brought to the brutal truth that she murders innocent people. It is this roller coaster, this depiction of Eli as complicated, that propels a film like *Let the Right One In* to horror movie legend.

CHAPTER FOURTEEN
BEETLEJUICE

Year of Release: 1988	
Director: Tim Burton	
Writer: Michael McDowell, Warren Skaaren	
Starring: Geena Davis, Winona Ryder	
Budget: $15 million	
Box Office: $73.7 million	

Beetlejuice is one of the movies that shaped us as children. I (Meg) became instantly drawn to Lydia (Winona Ryder). I dressed as her and even recently found a photograph of my *Beetlejuice*-themed eighth birthday cake. My mother created a Lydia poster for my room after weeks of my begging. Kelly remembers watching the movie whenever she could, wishing she could be friends with Lydia and secretly rooting for her to marry Beetlejuice (Michael Keaton). She realizes now that he may not have been the best choice for a life (or death?) partner, but she always had an affinity toward monsters!

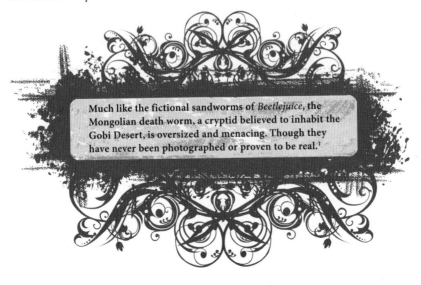

Much like the fictional sandworms of *Beetlejuice*, the Mongolian death worm, a cryptid believed to inhabit the Gobi Desert, is oversized and menacing. Though they have never been photographed or proven to be real.[1]

It would be easy to call *Beetlejuice* a wacky horror comedy and leave it at that. Beetlejuice himself is a misogynistic dead guy, who thankfully gets his comeuppance at the end of the film when he messes with the wrong Voodoo Doctor. And there is no doubt that the character of Beetlejuice provides a humor that charmed audiences upon its release in 1988 as it still does today. But it is Lydia who is the bleeding, gothic heart of *Beetlejuice*. In 2019, the musical version of *Beetlejuice* opened to sold-out crowds and was nominated for eight Tony Awards. For many girls, us included, Lydia represents a kind of teenager we hadn't seen before. A girl who likes the "strange and unusual" but is also kind, sweet, and holds a childlike innocence. Similar to *Let the Right One In*'s Eli, Lydia is not so simple as to exist in only one sphere. She is both light and dark, girl and woman.

Older than the creepy little girls of the previous chapter, Lydia is in the most prevalent age group of female horror movie characters. As a teenage girl, she joins everyone from Nancy (Heather Langenkamp) of 1984's *Nightmare on Elm Street*, to Sidney Prescott (Neve Campbell) of 1996's *Scream*, to pretty much every female populating the slasher genre. In fact, the slasher (defined by its antagonist hunting down a group of people with a sharp weapon) is synonymous with teenagedom, chronicling the trials and tribulations of both boys and girls as they travel the rocky road to adulthood. In a 2018 study, Kehksha and Deoshree Akhouri define adolescence:

[It] has been described as a phase of life that starts with biology and ends in society. Indeed, adolescence is the period within the life span when most of a person's biological, cognitive, psychological, and social characteristics are changing from what is typically considered childlike to what is considered adult-like. In this age, adolescents have to face various major changes and challenges required to adjust in the family, peer group, and society. These changes make them an active observer of the events happening around them. These changes often start from the elementary educational institutions to either junior high school or middle school. With the successively altering stages, they are expected to fit in all the settings properly by parents and society.[2]

Here again, we encounter society's expectations, the crushing narrative on adolescents to behave a certain way so as to please those around them. This battle, waged among nearly all teens against the world, is a well-used conflict in the horror genre.

Over $1 million of *Beetlejuice*'s budget was used for visual effects, including stop motion, prosthetic makeup, puppetry, and blue screen.

Lydia, unlike many of the teen girls populating modern horror, does not concern herself with boyfriends. Sex is naturally the most potent element of the transition to womanhood, but we covered all the complexities and nuances of that in Section Three (p. 45). Lydia is busy battling the stifling control of her stepmother, Delia (Catherine O'Hara), a woman who represents everything Lydia hates: greed, image, and insincerity. Lydia wants to keep their house as it is, charming, while Delia wants to destroy and renew it in her image. Home is considered a domestic, feminine place; therefore, the strain between Lydia and Delia for control over their new home is quintessential Female Gothic symbolism. Every girl must come to the crossroad at which she deviates from her maternal caregiver, or chooses to follow in her footsteps. Lydia is obviously one of the former teens, as even her exaggerated "goth" dress is a rebellion against Delia.

This leads us back to female rebellion. Teen girls are more likely to subvert expectations than those women twice their age. By simply being a teenager, it is their purpose. This usually manifests (both in the real world and in film) in sex and underage drinking (*Friday the 13th*, *Halloween*) or, in Lydia's case, embracing what makes her unique in the face of her critical stepmother. Yet, the purpose of horror, and fiction at large, is to examine what happens when these tropes are exaggerated. What happens when this female teenage rebellion takes a darker turn, creating a world of chaos and terror?

Truth is often a powerful catalyst for storytelling. *Lord of the Rings* (2001) director Peter Jackson and writer Fran Walsh mined the true story of the Parker-Hulme murder for their film *Heavenly Creatures* (1994). In June of 1954, Honorah Parker was killed by her sixteen-year-old daughter, Pauline, and her daughter's fifteen-year-old friend, Juliet. At a time before films like *The Bad Seed* even pointed to such girl-driven horror, it was a shock for the community of Christchurch, New Zealand. The 1950s was a time of convention, of a cohesiveness that all teen girls must act and look a certain way. Certainly, a girl dressed like Lydia, all black with a veiled hat, would be shunned and misunderstood at the time of Parker's death.

It is believed that one motive of the murder (the teen girls bludgeoned Honorah Parker with a piece of brick inside a stocking) was that the parents suspected Pauline and Juliet of having a romantic relationship. Homosexuality was strictly taboo in the 1950s. *Heavenly Creatures* is a psychological drama that focuses more on the relationship of the two girls (Kate Winslet and Melanie Lynskey in their film debuts) than the murder, perhaps finding sympathy in their story. It is a reflection on the loss of innocence, expectation, and girlhood. Totally different is the dark comedy *Heathers* (1988), despite dealing

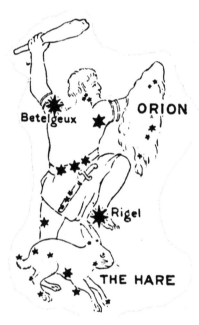

Betelgeuse is the ninth brightest star in the night sky, and the second brightest in the constellation Orion.[3]

with the same subject matter. Adam Markovitz described the quirky film for *Entertainment Weekly*:

In 1986, when John Hughes was giving us teen classics such as *Ferris Bueller's Day Off* and *Pretty in Pink*, (Daniel) Waters began working on what he calls "a Carson McCullers style novel of a girl who meets the Antichrist as a teenager." The project morphed into a screenplay about an angsty popular girl, Veronica (Winona Ryder), who starts secretly killing her classmates with the help of a diabolical new kid, J.D. (Christian Slater), and framing the deaths as suicides.[4]

Considered one of the best films of its era, *Heathers* plays upon our shared fear of the female teen, that she is not the helpless ingenue of slasher fare, but, like the creepy little girl trope, has the ability to cause destruction. As evidenced by the Parker-Hulme murder, as well as real-life fodder used for entertainment like *Willing to Kill: The Texas Cheerleader Story* (1992), female teen murderers are not a fictional premise.

Not all archetypes allow for such lateral growth. The girls of *Heathers* and *Beetlejuice*'s Lydia are created with wide strokes, flourishing under the understanding that characters should embody more than one attribute. Unfortunately, there are many teen girl characters who are given far less consideration. On the list of "Six Archetypal Horror Characters and Why They're Important," the archetype of cheerleader is very bluntly stated as:

Just the jock in female form. Not necessarily a cheerleader, she is a pretty girl who is not very bright and not very nice. She is usually the main female character's best friend, for reasons unknown to the rest of us. She is mean, but she is also beautiful so the other characters, mainly the guys, tolerate her. She is the object of desire or envy respectively, and is easy for viewers to hate and the villain to kill. This is always the character who becomes the most terrified in her situation. She becomes absolutely hysterical, which means she also becomes incredibly annoying. It's usually at this point that the viewer will start screaming, "Just kill her already!"[5]

Let us unpack this extremely misogynistic description of the cheerleader trope. Examples of this archetype would be Lynda (P.J. Soles) in *Halloween*

(1978) or Paige (Paris Hilton) in the remake of *House of Wax* (2005). First, we must point to the term *hysterical* used in a negative context. It is true that this archetype tends to be the most emotional, often allowing her emotions to cloud her logic, unlike the final girl. What's vital to note is that this assumed "hysteria" is deemed "annoying" by both the writer of the article and, often, much of the film's audience. It seems as if this cheerleader were asking for death by having a negative attitude, low intelligence, or high emotion, the latter of which is both understandable in the horrific scenario and almost always attributed to the female. In the study "Sex, Violence and Victimization in Slasher Films," in which thirty slasher horror films were analyzed, a fascinating finding came to the surface. Men were in greater number the victim (two hundred and twenty to one hundred and forty-six), yet women's fear was depicted more:

> The sample of slasher films contains an average of 679.8 seconds of perceived threat per film. A significant main effect for gender was found with females shown in fear longer than males. Using the Bonferroni method of multiple comparisons, it was found that the average number of seconds of threats directed at males was significantly shorter than threats directed at females.[6]

Those few years we are teenagers seem to stretch before us, often punctuated by exciting rites of passage that have been grist for the horror mill, like *Prom Night* (1980), *The Babysitter* (2017), and many more. Lydia of *Beetlejuice* remains a teen horror icon, a goth girl with heart who endures beyond a single film.

CHAPTER FIFTEEN

WHAT EVER HAPPENED TO BABY JANE?

Year of Release: 1962	
Director: Robert Aldrich	
Writer: Lukas Heller	
Starring: Bette Davis, Joan Crawford	
Budget: $980,000	
Box Office: $9 million	

By society's standards, little girls and teens are supposed to maintain a certain level of innocence. The truth is, women, too, have long been expected to preserve a demure exterior. In the beginning of 1962's *What Ever Happened to Baby Jane?*, we are introduced to poised and adorable nine-year-old Baby Jane (Julie Allred). She is the quintessential little girl, dressed in lace and blonde ringlets. After she performs a treacly song to an adoring audience, we witness Jane's spoiled behavior behind the curtain. Although her fans are buying dolls in her likeness, Jane is cruel in reality. This dichotomy, as discussed previously, is an unnerving display of what we expect and what can often be the horrific truth. Put girls, or women, in frilly frocks and have us dance, but that will not alter our true selves.

Time goes on, and next we see that older Jane (Bette Davis) is making an attempt at being a film star in adulthood. Unfortunately for Jane, her sister, Blanche (Joan Crawford), is much more successful at this, taking Hollywood by storm and leaving her formerly famous sibling in the dust. Something that seems vital to note is that in a modern rewatch of *What Ever Happened to Baby Jane?* it is men who make the decision on whether Jane or Blanche is talented. We watch as varying groups of men discuss their merits. Perhaps this is simply a product of the time, or could it be a purposeful action to showcase the helplessness of women in an industry led by men? Whatever the case, the film was both about its characters and the actresses portraying them:

> Although Jane seems to be the culpable one in the film, *What Ever Happened to Baby Jane?* implicates the spectator in each woman's self-destruction. We watch as two stars, churned out by the Old Hollywood system, tear the fine patina of glamour and vanity off. There's a point in the film where we stop watching Blanche and Jane and see Davis and Crawford in their "gothic grotesquerie." Each woman is vulnerable, shedding the image that each took so long to construct and maintain.[1]

Bette Davis created her own memorable makeup for the film, which included an over-the-top white-powdered face and thick, exaggerated lips.[2]

As Blanche and Joan age, the titular question is answered. What happened to Jane is that she is a drunk, still dressed in her Baby Jane attire and wishing life had been different. She has been used by the system she wanted so desperately to be a part of. Her desperation has seeped indelibly into her relationship with her sister, creating a venomous jealousy that causes Jane to control paralyzed Blanche and treat her with a cruelty more terrifying than little Jane could muster. Blanche has become a prisoner, in mental anguish because of Jane. We watch their toxic relationship crumble from the inside out, all the time wondering why they've let Hollywood, and society, dictate their lives.

In a time when movies barely passed the Bechdel test, in which two women have a conversation about anything other than a man, *What Ever Happened to Baby Jane?* dared to question the treatment of women both young and old in a male-driven world. Later in the film, Jane physically abuses her sister and even kills her caretaker Elvira (Maidie Norman). These extreme actions lead us to explore more women whose innocence has been eroded by society, leaving her to strike out murderously. This trope often begins with a woman who was mistreated in her childhood. Jane, for example, had the heavy burden of child stardom on her little shoulders, and the subsequent failures in contrast to her sister's successes. This leads to her brutal action, along with what Blanche and others believe is a mental break.

In more extreme cases, women who have endured bad childhoods or past events lose their innocence, becoming callous or even evil. This is in contrast to the wounded healer who uses her past trauma to navigate and ultimately win against the antagonist. Women like Jane, the ones who have twisted their victimhood, are a different breed. These women populate the revenge subgenre (see Section Four on p. 69) but this revenge against abuse often spirals out of control and hurts those who do not deserve the retaliation.

In *May* (2002), we are introduced to twenty-eight-year-old May (Angela Bettis). May had a strained childhood punctuated with bullying because of her lazy eye. Because of this, May does not function well socially as an adult and has no friends other than a doll given to her by her mother. This juxtaposition of childhood totems and adult arrested development is also evident in *What Ever Happened to Baby Jane?*.

People start to talk to May when her eye is fixed, but despite this apparent boost of "normal" physical attractiveness, May's disturbed psyche has not improved. She strikes out against those who've attempted to socialize with her, starting with the killing of her cat and then ramping up into serial murder. Like in the case of Jane, May is a victim of her circumstance, one who has allowed her trauma to lead her down a path of violence. Her innocence gone, May walks a tenuous rope of both clinging to childhood and embracing the horrors of adulthood.

In "What's Natural About Killing? Gender, Copycat Violence and *Natural Born Killers*," Karen Boyle discusses a similar loss of innocence in the film *Natural Born Killers* (1994):

> The transformation of the passive, victimized Mallory Wilson (Juliette Lewis) into the murderous Mallory Knox is revealed in the sitcom parody *I Love Mallory*. The sitcom is a flashback of Mickey (Woody Harrelson) and Mallory's first meeting in which Mickey, the star of the show, rescues Mallory from her sexually abusive father. Positioned near the beginning of the film, Mallory's subsequent violence is, therefore, read in the light of her initial victimization. This appears to have been a conscious strategy on the part of (director) Oliver Stone and Juliette Lewis. Lewis claims to have suggested this addition to Quentin Tarantino's original story: "I mentioned that (Stone) might wanna show that something happened to this girl in her background. It's hard to see a girl be that cruel. I didn't want to disgust the audience. I want them to understand the character a little." Thus, while Mickey's background, which is not revealed until much later in the film, confirms the inevitability of his violence, *I Love Mallory* emphasizes the reactive nature of Mallory's crimes. Sitcom Mallory is sexually abused by her father, her mother condones the abuse, and the complicit "audience" laughs along.[3]

When Juliette Lewis applies the word *disgust* to a female killer, she is pointing to the societal expectations of females and the expectation that if they are killers, there should be a loss of innocence, a victimization that doesn't excuse her deeds but does make them more palatable. This leads us back to Jane and Blanche at the end of *What Ever Happened to Baby Jane?*. They finish the film on a sunny beach in Los Angeles, a drastic change from their gothic mansion. Blanche lies dying in the sand. She feels the need to confess to Jane that she is the cause of the car accident that caused her paralysis, not Jane, who has spent her life in guilt over the act. "You were not ugly then. I made you that way." Blanche explains how she allowed everyone to blame Jane. The jealousy between them flowed both ways.

Sibling rivalry is more common among children who are the same gender and close in age.[4]

Jane can't handle this confession from her sister. She has painted herself as an attempted murderess, and there is no escaping this stigma. She has worn it as she's worn her thick Baby Jane makeup. It will simply not slip off. The innocence was gone long, long ago.

There are many themes at play in *What Ever Happened to Baby Jane?* including its critical look at the entertainment industry. One aspect that has changed drastically since Baby Jane sang "I've Written a Letter to Daddy" for her admirers in 1917 is the advent of social media and fan conventions. This tenuous relationship between fan and performer is explored in the upcoming film *13 Fanboy* starring Corey Feldman, Dee Wallace, and many more horror movie legends. The premise, that former actresses in the *Friday the 13th* franchise are being stalked by a fan who doesn't differentiate between fact and fiction, makes for a meta-horror film. We had the great fortune to discuss *13 Fanboy* with writer-director Deborah Voorhees, who also appeared as Tina in *Friday the 13th: A New Beginning* (1985).

Meg: **"First, tell us about Voorhees Films. What are the goals of your production company?"**

Deborah Voorhees: "I want to make all types of films at Voorhees Films. I love variety. Right now, we are focused on a horror thriller called *13 Fanboy*. It's about an obsessed fan stalking his favorite horror actresses from *Friday the 13th* with the intent to kill. We

have a stellar cast. Dee Wallace is our lead, along with newcomer Hayley Greenbauer. Our supporting cast includes many *Friday* alums including Adrienne King, *Friday*'s first final girl; Lar Park Lincoln, the final girl from Part 7; Kane Hodder, who played Jason more than any other stuntman; C.J. Graham, a favorite Jason from part VI; Thom Mathews, Judie Aronson, Tracie Savage, Ron Sloan, Jennifer Banko, and cameos from Carol Locatell and myself."
Meg: "**That sounds like a lot of fun!**"

Kelly: "**True crime has often been inspiration for horror films. Because of your history as a journalist, you were exposed to real-world horrors. Did these experiences inform the sort of stories you tell?**"

Deborah Voorhees: "*13 Fanboy,* while it is a fictional piece, it has some truth to it. I was surprised to see how many women from *Friday the 13th* have been stalked. I'm not talking about an ex being a nuisance or someone inappropriately bothering someone. These are very serious situations where someone was shot at, a stalker was living in someone's attic, and much more."

Meg: "**In *13 Fanboy* you explore the dangers of fame in a very 'meta' way. Could you give us insight into what it's like for a horror actress in the convention sphere? And do you believe it differs from male performers' experiences?**"

Deborah Voorhees: "That's an interesting question. Men, of course, are stalked too, but women have this issue more often and more often the stalking tends to be more serious. Adrienne King's stalker followed her to London. Lar Park Lincoln's stalker stalked her for six years and lived in her attic. She'd come home to find her newly cleaned nightgown neatly folded on her pillow or her furniture rearranged. Judie Aronson's stalker would leave letters and photos taken of Judie at her home, at her friend's homes, and such."
Meg: "**That is so scary!**"

Kelly: **"What sort of female characters intrigue you as a creator? Do you like strong, 'final' girls? Complicated antiheroes? And is it a priority of yours to tell female-driven stories?"**

Deborah Voorhees: "I love strong women who can kick butt. If my husband wants to watch an action adventure movie, I grumble. I don't need to see another one of those movies. But the second he tells me that a woman is kicking ass I yell out, 'grab the popcorn.' I love women in action, taking charge. Now, given all that, my next statement may sound strange. I try to not focus on 'female'-driven stories even though I very often write stories for strong women leads. We never say male-driven story. We assume that our strong lead is male. We have to get to the point where we stop talking about putting women in front because it is the right thing to do, and simply put women in front because they make for great stories, and let that be the norm."

Deborah Voorhees gave us incredible insight into the world of women in entertainment. While *What Ever Happened to Baby Jane?* is a striking portrait of women's innocence stripped away over the decades, we look forward to seeing the reality-infused *13 Fanboy*, developed by a woman who has lived in the industry.

SECTION SIX
THE GORGON

CHAPTER SIXTEEN

DRAG ME TO HELL

Year of Release: 2009	
Director: Sam Raimi	
Writer: Ivan Raimi, Sam Raimi	
Starring: Alison Lohman, Lorna Raver	
Budget: $30 million	
Box Office: $90.8 million	

"Everything written for women seems to fall into just three categories: ingénues, mothers, or Gorgons." Jessica Lange says this line as Joan Crawford in *Feud* (2017) and speaks to a time in Hollywood that was, perhaps, even more limiting for women than it is now. Roles for women seem to diminish after a certain age, and if available, the parts paint women into

Medusa is known as one of the Gorgons.

very specific types. One of these types is the concept of the Gorgon, which dates back centuries to Medusa. The Roman poet Ovid wrote of her transformation in *Metamorphoses* (8 AD). Medusa was once a beautiful young maiden, the only mortal of three sisters known as the Gorgons. Her beauty caught the eye of the sea god Poseidon, who raped her in the sacred temple of Athena. Furious at the desecration of her temple, Athena transformed Medusa into a snake-haired monster with the ability to turn whoever looked upon her face to stone. This seems like a setup for a unique, redemptive, female heroine. A man becomes the hero of the tale next, though, as he is sent to slay Medusa and use her severed head as a weapon.[1] Talk about stealing the climax from the woman! "In Western culture, strong women have historically been imagined as threats requiring male conquest and control, and Medusa herself has long been the go-to figure for those seeking to demonize female authority."[2]

Studies show that how a group of people are portrayed in media can have an effect on how those people are viewed in the real world. This rings true for older women. Ageism is discrimination based on a person's age and often affects the elderly. When older women with power or authority are portrayed in literature or film, they are sometimes shown as shrewd, scary, or unlikable. A study in 1993 found that "the quality of media representation of the elderly was poor and inadequate. Elderly characters both in television and print media were typically marginalized. They rarely appeared in major roles or positions, were rarely developed fully as characters, and were frequently described in stereotypical terms."[3]

Some of these stereotypes include showing older people in poor health, being hard of hearing, forgetful, and having few personal relationships. Other older females are often shown in minor background roles such as grandparents or neighbors and are not always fully fleshed-out characters. In 1990, it was found that only 3.3 percent of roles on television were portrayed by women over the age of sixty-five. In 2018, those same roles for women made up a mere 1 percent on television.[4] How does the horror genre compare in representing older women? The roles for older women in horror date back to its inception, but most notably, a certain type of role for older women came about in 1962 with the film *What Ever Happened to Baby Jane?*. "Hagsploitation" or "psycho biddy" are the terms used to describe

Psycho biddy roles explored previously glamorous women later in life.

the subgenre of horror that explores how a formerly glamorous woman has become older, unstable, and terrorizes those around her. This premise is further explored in *Hush Hush Sweet Charlotte* (1964), *The Nanny* (1965), and *Berserk* (1967). These movies offered older women roles but weren't necessarily the most flattering portrayals.

Drag Me to Hell (2009) features an older female character in Sylvia Ganush (Lorna Raver), an elderly woman who is looking for some more time to pay on a loan through her bank. When the institution denies her request, she places a curse on the loan officer, Christine, played by Alison Lohman. On her character's motivations, Raver said, "I see Sylvia as a woman who justifies her actions because she comes from an old-world culture. She reacts viscerally because of who she is. Maybe she takes it too far, but she certainly does have cause for complaint with how Christine treats her."[5] Raver is an actress who has performed in plays across the United States and eventually began making movies in Hollywood: "I haven't done a lot of film work because it's a difficult area for an older woman of a certain age who is not a big name." Speaking of being cast in *Drag Me to Hell*, Raver said, "When this came up and I found out that Sam [Raimi] was involved in the project, I was very excited."[6]

A "lamia," on which Mrs. Ganush is based, is a bogeywoman from Greek mythology who stalks the countryside looking for children to devour; it was a story used to get young children into their beds at bedtime.[7]

Raver auditioned for the role and saw only small pieces of the script, which is typical for this situation. "I had no idea what I was getting into, because all I had read was about a little old lady coming into the bank because they're closing down her house. It was only later that I saw the whole script and said, 'Oh my!'" The movie was directed by Sam Raimi, who began his career directing horror movies such as *Evil Dead* (1981) and *Evil Dead II* (1987). Raver said, "Sam has these touches that are a little bit off-center that break the tension. He's great to work with as an actor because he includes you in the process. I found it interesting to watch him on the set because he's very focused, and sometimes you can see the movie running behind his eyes." Even though this was Raver's first major film role, Raimi was impressed with her: "Lorna went to town with this role, especially in the car attack scene. She's a real fighter who was always willing to give you one more take and put everything she had into it."[8]

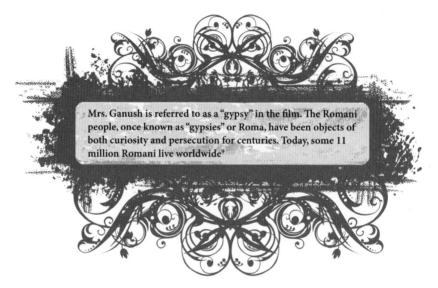

Mrs. Ganush is referred to as a "gypsy" in the film. The Romani people, once known as "gypsies" or Roma, have been objects of both curiosity and persecution for centuries. Today, some 11 million Romani live worldwide[9]

Raver also had to work closely with KNB Effects Group, which she said turned out to be the hardest part of being in *Drag Me to Hell*. Raver said:

They had to do a live cast of my entire body for the movie so that they could build puppets and work on some other makeup effects. It was five guys working all at once, but it still terrified me because I am claustrophobic. I was so nervous, but I knew I was in good hands with them. They did such an amazing job that when they asked me to do a second live casting with me screaming, I couldn't say no to them."

The Gorgon character in *Drag Me to Hell* follows the trope of an older woman having magical or otherworldly powers. This can be seen in *The*

Skeleton Key (2005) and the *Insidious* film series (2010–2018). How have women, in general, been portrayed in media and literature? According to author Anita Nair, "Literature has always been ambivalent in its representation of women. Good women, as in ones who accepted societal norms, were rewarded with happily ever after. Even feisty heroines eventually go on to find content and life's purpose in a good man's arms . . ."[10] Until recently, males were writing and creating female characters who were often shown as a virgin or a temptress. With more diverse creators of literature, film, and television, we can only expect to see people of all ages, backgrounds, and types be represented. And we are here for it.

CHAPTER SEVENTEEN

THE VISIT

Year of Release: 2015	
Director: M. Night Shyamalan	
Writer: M. Night Shyamalan	
Starring: Olivia DeJonge, Deanna Dunagan	
Budget: $5 million	
Box Office: $98.5 million	

Staying with relatives can sometimes be a nightmare, but for the children in *The Visit* (2015), their nightmare becomes a reality. *The Visit* is a found footage–style movie that follows two teenagers, Becca (Olivia DeJonge) and Tyler (Ed Oxenbould), as they stay with their grandparents for a week. They've never met their grandparents due to their mother (Kathryn Hahn) being estranged from them. What could go wrong? Plenty. It's the perfect setup for a horror movie plot filled with surprising twists and turns.

Speaking about her character in *The Visit*, Deanna Dunagan said Nana is:

Not exactly what you first think when you initially see her in the movie. I don't want this to sound very grand and all, but I think it's true that I thought of Nana the same way I do about [Eugene] O'Neill's characters. O'Neill's characters often turn quickly from one mood to another, they turn on a dime. That was also true of Nana in *The Visit*. So, that was what was so great about playing her.[1]

Child actors, like the ones in *The Visit*, have numerous laws for their protection while on film sets including time limitations, procedures, use of stunt doubles, and schooling requirements.[2]

The grandmother in the film is said to be suffering from sundowning condition. The term *sundowning* refers to a state of confusion occurring in the late afternoon and spanning into the night. It can cause a variety of behaviors such as confusion, anxiety, aggression, or ignoring directions. It can also lead to pacing or wandering. While the exact cause of this behavior is unknown, it typically affects those with dementia and Alzheimer's. This condition, and other symptoms portrayed in the film, play into our society's fear of aging. As people are living longer, we are seeing their health deteriorate in ways previous generations may not have. The children in *The Visit* are advised that their bedtime will be 9:30 p.m. while staying with Nana and Pop Pop (Peter McRobbie). They don't outwardly protest but are bothered by the prospect. On the first night, Becca begins to sneak downstairs to get a snack. She is shocked by the sight of her grandmother, wandering about and vomiting at the foot of the stairs. It's a strange and disturbing moment, but not completely out of the ordinary. It's explained away by her grandmother having a "touch of the stomach flu." The grandfather has a shed filled with his adult diapers that reek of human excrement. This is explained away by him suffering from incontinence due to old age. These things could be commonplace but are used to make the grandchildren, and the audience, uncomfortable.

Another horror film that explores a real disorder affecting someone of old age is *The Taking of Deborah Logan* (2014). Once again in the style of a found footage movie, a film crew documents the title character, played by Jill Larson, as she succumbs to the throes of Alzheimer's disease. Her physical and mental condition deteriorate throughout the movie, and we are faced with supernatural elements that are contributing to her state. The moments of witnessing a daughter watch her mother lose sense of reality are heartbreaking and reminiscent of real grief. The story takes a dark turn, and Deborah is never able to recover from her condition.

There is a vast difference in how various cultures treat their elderly and think about their aging population. The elderly in Fiji, for example, are cared for by their families all the way through to death, while many elderly family members in the United States live out their days in a nursing home or other facility. Extreme examples of how the elderly have been treated in the past include nomadic groups of people leaving behind sick or older individuals in order to keep making progress throughout a move. Some societies even sacrificed their weakest members during famine to guarantee survival of the youngest or strongest. In modern cultures the traditions can vary widely. Many East Asian cultures believe in Confucian tradition that places a high value on filial piety, obedience, and respect. Researcher Jared Diamond said, "It is considered utterly despicable not to take care of your elderly parents." In many Mediterranean cultures, multigenerational families live together in the same house, in stark contrast to the United States, "where routinely, old people do not live with their children and it's a big hassle to take care of your parents even if you want to do it."[3]

Not all grannies are nice! Nannie Doss was an American serial killer responsible for the deaths of eleven people between the 1920s and 1954. Nannie Doss was referred to as the Giggling Granny.[4]

The topic of aging is explored in 2019's *Midsommar*. A group of Americans travel to Sweden and witness a culture that believes when you turn seventy-two you should choose to die, on your own terms. This practice exists in some cultures and is called "voluntary death." The group in the film sees older people purposely take their own lives in order to avoid becoming a burden on others and to evade potential health problems later in life. The outside group finds the practice horrific but doesn't feel that they have the power to influence another culture's beliefs. According to a study in *American Ethnologist*, the Chukchi of Siberia practice voluntary death, in which an old person requests to die at the hands of a close relative when they are no longer in good health.[5] Norse tribes in Scandinavia follow similar practices. The elderly will enter into a difficult situation, like setting out to sea on a solo voyage, in order to end their lives. The Ache of Paraguay let the elderly men in their culture wander off to die and will kill elderly women by breaking their necks.[6]

The family in *The Visit* isn't confronting voluntary death or end of life care, but they are estranged. Studies show that estrangement itself is not all that uncommon. One study of adult children in the United States found that 7 percent reported being detached from their mother and 27 percent detached from their father. These detached relationships were characterized by infrequent or no contact or support, feeling distant from the parent, having different values from those of the parent, and rating family as a low priority. A German study found that over 10 percent of adults over the age of forty reported intergenerational family conflict, and half of these stated that they avoided the other person or had ceased contact altogether. There is a stigma associated with reporting estrangement and a difficulty in defining it, so it is likely that the rates are much higher than people are reporting. Additionally, these studies do not account for the estrangements that occur between siblings, grandchildren, and other extended family.[7]

Women have always been featured prominently in fairy tales and are often portrayed as virtuous maidens but also as witches or evil stepmothers. Dr. Silima Nanda notes that "fairy tales embody the ways that societies attempted to silence and oppress women by making them passive. Much of the fairy tale literature reinforces the idea that women should be wives and mothers, submissive and self-sacrificing. Good women in stories are to be silent, passive, without ambition, beautiful, and eager to marry."[8]

A scene in *The Visit* is reminiscent of the fairytale *Hansel and Gretel* (1812). Nana gets Becca to go into the oven by having her help clean it. With the added tension of camera placement, we can't entirely see if Becca will be safe or not, and it brings the scene to another level. Like the old woman in *Hansel and Gretel*, we are at first unaware of Nana's true intentions. In "Why Are Old Woman Often the Face of Evil in Fairy Tales and Folklore?" Blair Elizabeth describes old women villains as:

> Especially scary because, historically, the most powerful person in a child's life was the mother. Children do have a way of splitting the mother figure into . . . the evil mother—who's always making rules and regulations, policing your behavior, getting angry at you—and then the benevolent nurturer—the one who is giving and protects you, makes sure that you survive.[9]

This dichotomy is explored in horror literature and films and is one that many people can relate to. In Russian folklore, Baba Yaga is a fairy tale that fits into the Gorgon myth. The first written reference to her was in 1755 in Mikhail W. Lomonosov's *Russian Grammar*, where she is listed among ancient figures from Slavic tradition. This legend likely began in pre-Christian, pagan times. The Russian witch is described as a deformed, scraggly old woman with bony legs, a long, crooked nose, piercing cold eyes, and iron teeth. Her demeanor is powerful and unpredictable. Her intention is to instill fear and respect in anyone who encounters her. Each body part of the Baba witch is described as grotesque. She lives deep in the forest in a hut that rests on giant chicken legs that can move around the forest to make it harder for anyone to find her. Baba's ambiguousness, according to the folklorist Joanna Hubbs, is "directly connected to her femininity, and her femininity to the natural world. Baba Yaga is an aspect of a great mother goddess, whose dual nature as genitrix and cannibal witch reflects a fundamental paradox of nature. In some ways, she's an earth mother figure; in others, she's closely associated with death."[10]

In Japanese folklore, Yama Uba is an old crone who lives and hunts in the mountains and eats anyone who crosses her path. She will often pose as a young woman offering shelter to lost travelers. Once the traveler falls asleep she will eat them, sometimes using her hair to trap her victims, and drag them into her enormous mouth. It is said that the Yama Uba was a normal woman until the area where she was living experienced a famine. Her family couldn't feed her and drove her out into the woods to starve. She eventually found shelter in a cave but was driven mad and started to feed off of people, turning into the Yama Uba from desperation and rage.[11]

These Gorgon characters from folklore sound like they were written to be villains in horror movies like *The Visit*. Several scenes in the film are absolutely terrifying, and they all have one thing in common: they feature the grandmother. The first scene that received shocked screams in the theater was an innocent hide-and-seek chase sequence beneath the house. We see Nana, who is usually

Older women are often portrayed as villains in fairy tales.

sporting a formal updo, with her hair down over her face. "Here I come!" Nana gallops in a crawl past Becca, and we are treated to two more jump scares before leaving the space. The grandmother emerges with a torn dress but doesn't seem concerned. The confusion and terror for the grandchildren is clear, though, and the audience shares their uneasiness.

The next horror-filled sequence featuring the grandmother comes on the second night of the visit. The children hear a strange, scraping sound coming from outside their bedroom door. They open the door to reveal their grandmother, completely naked, scratching at the wall. These strange, disturbing behaviors may seem contrived for a horror movie, but for those who have watched relatives fall into the clutches of forgetfulness or dementia, they are all too real. Becca tells her brother, "Just come to accept that they're old people and things won't be as weird." Their mother explains it even more simply: "They're old."

On the third night, the children discuss setting up a hidden camera in the living room to catch any strange "sundowning" behavior that their grandmother may be partaking in. After debating the moral implications of filming someone without their knowledge, they decide not to. That night we see, from the perspective of their handheld camera, Nana running back and forth across the hallway with her hands held behind her back. Again, this is a potentially benign act, but one that instills fear in the casual observer. When we watch from the position of the hidden camera on the following night, a jump scare occurs that caused moviegoers to scream and leap in their seats. On this night, Nana is wandering around the living room and pops up, stares into the camera, and growls. She carries the camera with her upstairs, and we see her try to enter the children's bedroom, knife in hand.

By the time we get to the last night of the children's stay, the audience has witnessed escalating strange behavior, and we have discovered, with Becca and Tyler, that the people they are staying with are not their grandparents. Nana stares blankly and frozen as Tyler sits awkwardly. A later scene follows the viewpoint of Becca and her single camera light following the woman claiming to be Nana around a darkened bedroom. The woman cackles, grabs blindly from under the bed, and moves with unexpected quickness. As we watched this scene in the movie theater for the first time with fellow moviegoers, the anticipation and fear was palpable. A physical confrontation concludes the scene, and we see the Gorgon defeated.

Older female characters are often used to incite fear in the audience. The character of Mrs. Kersch (Joan Gregson) in *It Chapter Two* (2019) was featured in the trailer for the film. An extended scene showed the elderly woman acting a bit strange, then moving in an inhuman way, and finally culminates with her appearing stark naked and confronting Beverly (Jessica Chastain). *The Shining* (1980) contains the iconic scene of the old woman in room 237, Mrs. Lorraine Massey (Lia Beldam), emerging from the bathtub as a young and beautiful woman, only to turn into an old woman covered in sores. The character has had numerous interpretations and theories surrounding her, but as Courtney Enlow said in an article for SYFY wire:

What she also is, is the hero of the movie as far as I'm concerned. She's the only one getting anything done around here. Everyone else either dies or experiences a dangerous mental breakdown or has lengthy freak-out sessions. In one scene, Mrs. Massey manages to enjoy a luxurious bath, make out with a young Jack Nicholson, and then terrify him and destroy his very mind for the duration of the film, and her time onscreen ends with a hearty laugh, unashamed of her body in all its forms. Quality work, Lorraine; well done.[12]

May we all have the unabashed confidence of Lorraine Massey in our later years—or at any age!

CHAPTER EIGHTEEN
FRIDAY THE 13TH

Year of Release: 1980	
Director: Sean S. Cunningham	
Writer: Victor Miller	
Starring: Betsy Palmer, Adrienne King	
Budget: $550,000	
Box Office: $59.8 million	

Pamela Voorhees, played by Betsy Palmer in *Friday the 13th* (1980), could fit into several sections of this book. She is, importantly, a mother. Her son, Jason (Ari Lehman), died of drowning while she was working at Camp Crystal Lake, a summer camp for kids. She could also fit into our section on revenge: she is, after all, seeking revenge on the counselors, and other perceived villains, who allowed her son to perish. The entire movie, and subsequent franchise, is built on the fear of Jason Voorhees being the killer. It is revealed, of course, that Jason was never alive in the original movie and that it was his mother who was the villain all along. Pamela is an older woman and presumably innocuous. We don't automatically see motherly, or grandmotherly, types as villains. This is what makes the reveal at the end of the first *Friday the 13th* movie such a gut punch: "Kill them, mommy!" When Palmer speaks this line in a voice other than her own and embodies her dead son, the audience feels palpable fear.

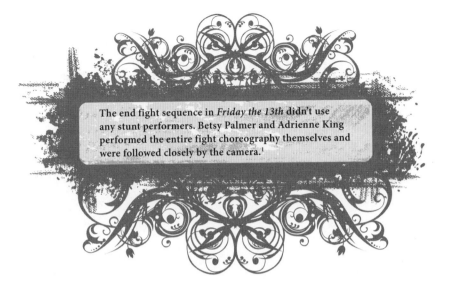

The end fight sequence in *Friday the 13th* didn't use any stunt performers. Betsy Palmer and Adrienne King performed the entire fight choreography themselves and were followed closely by the camera.[1]

Palmer was a theatrically trained actress who performed in plays, television, and films throughout her career. Her credits include *The Long Gray Line* (1955), directed by John Ford, and *Queen Bee* (1955), starring Joan Crawford. The famous story of how she took on the role in *Friday the 13th* starts with a broken-down car. Palmer was performing on Broadway in New York City and commuting to Connecticut when her car conked out. She found a replacement that she wanted to buy for $9,999 and was offered the role of Pamela Voorhees for $10,000. She didn't think anyone would ever see the movie, which she referred to as "a piece of junk," and agreed to take the role.

"Pamela serves as both Jason's strength and weakness over the course of the franchise; just as she killed for him in the first movie, he then goes on to kill for her. I've heard of the mother and son bond being strong, but these two take that to a whole new level."[2] Pamela was a single mother raising her son after an abusive relationship. According to the Centers for Disease Control and Prevention, nearly half of all women in the United States have experienced psychological aggression by an intimate partner in their lifetime, and 35 percent have experienced rape, physical violence, and/or stalking by an intimate partner.[3]

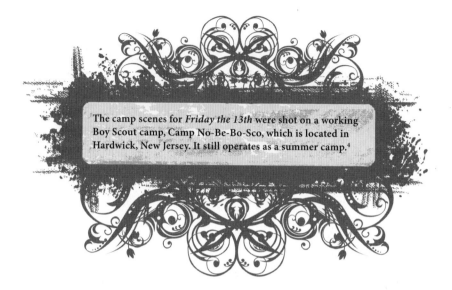

The camp scenes for *Friday the 13th* were shot on a working Boy Scout camp, Camp No-Be-Bo-Sco, which is located in Hardwick, New Jersey. It still operates as a summer camp.[4]

Jason's backstory includes his being born with the condition of hydrocephalus. The term *hydrocephalus* is derived from the Greek words *hydro*, meaning water, and *kephalon*, meaning head. The primary characteristic is excessive accumulation of fluid in the brain. The causes of hydrocephalus may result from inherited genetic abnormalities or developmental disorders. In infancy, the most obvious indication of

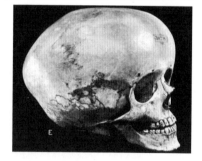

Hydrocephalus is characterized by head enlargement in infants.

hydrocephalus is a rapid increase in head circumference or an unusually large head size. Other symptoms may include vomiting, sleepiness, irritability, downward deviation of the eyes, and seizures. Older children and adults may experience different symptoms including problems with balance, slowing or loss of developmental progress, lethargy, drowsiness, irritability, or other changes in personality or cognition including memory loss.[5] Learning more about Pamela's backstory and knowing what she did to try to protect Jason in life doesn't excuse her behavior later on, but it does make her character more complicated and sympathetic. We don't know Pamela's exact experience in parenting a child with a medical condition, but over nine million Americans are actively caring for a child with special needs at any given time. According to the U.S. Department of Agriculture, 28 percent of children with disabilities are more likely to live in families that are considered poor, or below the federal poverty threshold.[6] With her theater training, Palmer created a biography for her character in *Friday the 13th*:

> I was taught the real method of acting where you do an autobiography for the character and you make a story happen before you come on stage. These people always have a life before you portray them. My story for myself about Pamela Voorhees was that she was in high school, and she and this boy fell in love and they were going steady. And in those days gals just didn't go to bed with guys. So I figured they did make love and she became pregnant. So she finally has to tell her parents and her father throws her out of the house. I had done a lot of work for the Salvation Army, they had a service for unwed mothers, so I figured she went there and had the baby.[7]

Unwed mothers are a theme in horror that has been explored in numerous movies including *Carrie* (1976) and *The Devil's Doorway* (2018). We spoke to Aislinn Clarke, the writer and director of *The Devil's Doorway* to ask her more about this theme.

Kelly: **"How did you explore this concept and the real history it was based on?"**

Aislinn Clarke: "Growing up in Ireland in the 1980s and 1990s, I was well aware of the Magdalene Laundries. These were church-run institutions in which undesirable women were housed, taken away from their families, and forced to work washing sheets and linens. A woman could be put in these places on the say-so of any male member of her family. They may have been lesbian women, mentally ill women, women that were seen as stubborn or standing in the way of male power or property. The vast majority of them were unwed mothers—women who would have their lives taken away as well as their children. While they were known by wider Irish society, they weren't really talked about. There was one not far from my home and my father used to deliver bread to it. He said it was like a vision of Hell—white hot hell—and that always stuck with me. My mother knew a girl who was taken off to a laundry when she was thirteen. The last one closed in 1996, the year before I had my son. I was seventeen at the time and, in a different family, could have easily been one of these girls. I have a lot of empathy for them and the ways in which their lives were given up to main the church-state apparatus in Ireland. I had done a lot of research for a documentary that was never made and then, when a producer came to me with an idea to make a horror film set in a Magdalene Laundry, I knew I had to do it. In recent years, the revelations of children's bodies found in the septic tanks of mother and baby homes in Ireland were horrific and, if they were going to be the basis for a horror film, it had to be done in such a way that the women and children at the heart of the story weren't exploited or demonized—they are the victims and the mechanism by which the church and state in tandem ran these place is the real horror. That is why I wanted the pregnancy of Kathleen—the 'possessed' girl—to be ambiguous: it could be an act of God or the Devil, it could ultimately be positive or negative for her, but the institution vilifies her for something beyond her control."

Meg: **"What has your experience been like as a woman working and creating in the genre of horror? Do you think the word is still seen as a pejorative?"**

Aislinn Clarke: "I don't think it is a pejorative, but it is certainly a label and there is the sense in some quarters that people follow the directions on the label very closely, that they consider the label rather than the artist. On the positive side, the Women in Horror community has certainly been very good to me and very welcoming and it is other female writers and directors—mostly, but not exclusively—that have been especially generous with their time, their advice, their support, and their camaraderie. There is a feeling that we are in it together and can help each other, which I'm not sure is necessarily the same experience that my male peers have. While my work has been well received generally, there is a section of academia specializing in horror from female filmmakers that has taken the work very seriously and really understands it. That pleases me immensely. I think it's important work. However, there is the risk in all this that female filmmakers get ghettoized as such, that their work is seen in some way as oppositional to the default, storytelling position, which is male filmmaking—women filmmakers being interpreted or evaluated on their difference or similarity to male filmmakers, rather than their work being considered in and of itself. This is the bind of the label: it is great to see more female filmmakers presenting work that incorporates a female point of view (while, of course, there are as many female points-of-view as there are females); however, women are then expected to tell women's stories, to fulfil the female quota—we haven't reached the point where women are given the benefit of the doubt as to whether or not they're artists. I want women's stories to be told, but, more than that, I want women to tell whatever stories they want. I think this block is apparent in the opportunities that female film-makers get. I am offered a lot of scripts with pregnancy and mothers, which are subjects I have an interest in because of my own experience of these things. However, I am not necessarily offered them because I, as an artist, am interested in them, but because —at this moment in time—producers know that such topics should be handled by women. It would be nice to reach a point where those same producers sent women scripts about war, about firefighting, about whale hunting, and other traditionally masculine topics in order to get a female POV on those things. Or, more importantly, that filmmaker's POV on them as an artist. In some instances, it is clear that the industry is interested in women because it knows it should be interested in women—thus they pay deference to the label. But the label is not there simply because woman is a category and that any woman will fill that role; the label is just a reminder that there are women and that women should be considered to fill any role."

Unwed mothers in *The Devil's Doorway* were forced into difficult circumstances, as was Pamela Voorhees in *Friday the 13th*. Betsy Palmer came around and eventually embraced her role as a horror icon. "Everybody wants a mother who will kill for you and will die for you—and I do both," she told the Newark *Star-Ledger* in 2005. An older mother seems harmless, and to have her revealed as the killer is one of the great plot twists in horror movie franchises. But older women are defying expectations offscreen, as well. According to MIT expert Joseph F. Coughlin, women are better prepared for life after middle age than their male peers.[8] He argues that women live longer and are more educated than any time in history. They also tend to be the primary caregivers, make purchasing decisions for their households, and do the research for most decisions for their families. Whether it be in real life or in the media, women having a more active role in society is long overdue.

SECTION SEVEN
THE HEALER

CHAPTER NINETEEN
STRANGER THINGS

Years of Production: 2016–present	
Created by: Matt Duffer, Ross Duffer	
Starring: Winona Ryder, Millie Bobby Brown	
Network: Netflix	

A phenomenon began in the summer of 2016: the first season of *Stranger Things* dropped on Netflix to immediate fanfare. The nostalgia-fueled TV series set social media ablaze with theories, memes, and die-hard fanship. One reason for the success of *Stranger Things* is undoubtedly its uncanny ability to hark back to the eighties' brand of horror, as it's an homage to everything from *Poltergeist* (1982) to *E.T.* (1984). But the true genius of the series comes in the form of well-developed characters for whom the audience can root. This includes a group of ragtag preteen boys, similar to *It*'s (1990, 2017) "Loser's Club." When Will (Noah Schnapp) mysteriously goes missing after a night of Dungeons and Dragons with his buddies Mike (Finn Wolfhard), Dustin (Gaten Matarazzo), and Lucas (Caleb McLaughlin), it is clear something supernatural is at play. This is emphasized when we meet a girl (Millie Bobby Brown), shaved bald and acting peculiar, who appears to have telekinetic abilities. This girl refers to herself as "Eleven," which we later learn is because she is merely a number in a lab, a test subject gone awry.

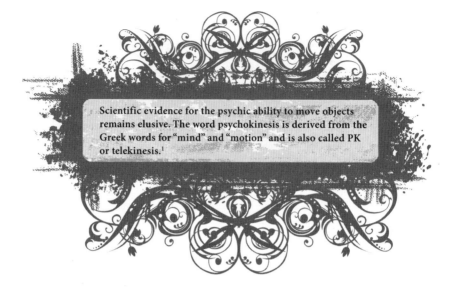

Scientific evidence for the psychic ability to move objects remains elusive. The word psychokinesis is derived from the Greek words for "mind" and "motion" and is also called PK or telekinesis.[1]

Eleven is arguably the heart of *Stranger Things*. She is a mystery all her own, with a troubling backstory of neglect, yet an ability to charm and show empathy. She wants to be loved and to be a "normal" girl. This archetype of a female equipped with special abilities that aid in the obstacles before her and those around her can be organized under the "healer" category. This supernatural sense is inherently of a healing variety: "They know things about the health and needs of the person without being fully aware how they know what they know. They have a sixth sense about what is going on in a person's being and a connection to the intuitive world of healing or knowing what that person needs."[2] While the boys she befriends hold no supernatural skills, Eleven is charged with more energy, more power, than she even knows how to control. She can flip vehicles, push humans, and in the end, she can save her friends and the town of Hawkins, Indiana, from the creatures of the "Upside Down."

This young girl with telekinetic capability is one trope familiar to the horror fan. In *Carrie* (1976), these skills could be used for good, yet Carrie (Sissy Spacek) is so damaged by her mother and her fellow students, she does not use her intuition to heal her wounds or those around her. Instead, she summons her powers as an act of revenge, killing everyone who has wronged her, including herself.

In *Firestarter* (1984), adolescent Charlie McGee (Drew Barrymore) has a set of circumstances similar to Eleven's. She is a girl with a scientifically altered intuition, one that starts fires and assists her in seeing the future. This makes her a sought-after commodity, leading to her kidnapping. Thankfully, Charlie is able to use her pyrokinesis for good, as she kills the bad guy, John Rainbird (George C. Scott), and presents the truth of the damaging scientific studies to the media. Eleven also exposes the truth, just simply with her existence, making both girls healers in their action to bring light to the dark goings-on of their communities.

In 1909, Stanisława Tomczyk purported to have similar powers as Eleven in *Stranger Things*.

In *Stranger Things*, Joyce (Winona Ryder) believes she is convening with her dead son, Will, only to discover he is in the "Upside Down," a place not unlike how death has been imagined. Joyce does not have the ability to speak to the dead, but this is a phenomenon often explored in horror. Mediums or spiritualists are an offshoot of the healer trope, and while not always female, women are often depicted as being closer to the spirit world.

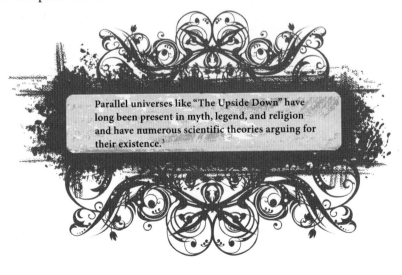

Parallel universes like "The Upside Down" have long been present in myth, legend, and religion and have numerous scientific theories arguing for their existence.[3]

Modern spiritualism is thought to have been born in 1848, when John D. Fox, his wife, and six children moved into a house in Hydesville, New York. Two of his daughters were purported to have found evidence of ghosts and further contended that they were able to communicate with these ghosts through a series of knocks. Their story was covered near and far, causing an eruption of spiritualism throughout the US. In the 1880s, Mary Fox admitted that she and her sister had fabricated the events that had led to America's fascination with spirit communication. She was quoted as saying, "There is no such thing as a spirit manifestation. That I have been mainly instrumental in perpetrating the fraud of spiritualism upon a too-confiding public many of you already know. It is the greatest sorrow of my life. . . . When I began this deception, I was too young to know right from wrong."[4] Later, Mary Fox recanted this confession, stating that her communications were actually authentic. Whatever the case, mediumship has stitched itself into the fabric of our modern world. Flip channels, and you'll find the reality show *Long Island Medium* (2011–present), then you may come across a rerun of the Patricia Arquette NBC drama *Medium* (2005–2011), or perhaps you fancy *Ghost Whisperer* (2005–2010)? Whatever your taste, there is a female spiritualist for you. Tangina (Zelda Rubinstein) of the *Poltergeist* films is a memorable medium, as she truly heals with her soft yet determined presence.

The healer myth exists in many feminine forms, including that of witch. Warlocks do have a place in modern myth, but witches have more successfully seeped into every kind of media. Their dominance reigned long before film: "Throughout the fifteenth century, the number of women tried for sorcery and witchcraft was significantly higher than the number of men."[5] A dubious honor to be sure.

Like many archetypes, the witch can be split into the good and the bad. Simply picture the Wicked Witch of the West (Margaret Hamilton) versus Glinda the Good Witch (Billie Burke). These two parallels from *The Wizard of Oz* (1939) provide startling insight. Glinda is the very image of goodness. She uses her witchy powers to help others, rather than for her own gain. This sort of witch provides far less fun in the horror genre, and so she is understandably scarce. A recent example would be Sabrina (Kiernan Shipka) in the Netflix series *Chilling Adventures of Sabrina* (2018–present), as she shows empathy, often employing her knowledge of the occult to help her human friends. One could argue this is because she is half-human. Samantha (Elizabeth Montgomery) of *Bewitched* (1964–1972), more comedy than horror, would also fall under this category. Even the aforementioned pure-hearted mediums or spiritualists could be considered within the healing witch archetype, as their powers are not entirely different. Centuries ago, a woman like *Poltergeist*'s Tangina would be burned at the stake.

More often, the witches who populate horror live up to their stereotype. Like the Wicked Witch of the West, they are self-interested: "The witch figure presents an awesome image of the primordial feminine concern with herself. Maternal life spends itself like life's blood flowing outward to nourish the sounds and bodies of loved ones. In the witch figure, life flows inward and downward to fuel the dark recesses of a woman's psyche or a man's anima."[6]

In other words, a witch is inherently dissimilar to the female role of mother. She allows herself to be selfish. Again, it seems that the expectations put upon a woman, to be the selfless Glinda, lead to the creation of the dark, inner Wicked Witch. A pervading theme in horror is that these hefty expectations, and the females' daring to rebel against them, create the sinister female archetypes from creepy girls to devious witches. In "Mothers, Witches, and the Power of Archetypes," Dale M. Kushner explains:

> The witch reminds us there may well be unnamable and untamable aspects of ourselves where passions stagnate and fester. What parts of us don't fit into the conventional idealized feminine? Do we harbor an urge to transgress and cross borders? Historically, innocent women have been tortured and killed because the prevailing masculine rule feared female sexuality.[7]

For his piece in *Essays in Medieval Studies*, Michael David Bailey further illuminates this point by referencing the famous witchcraft text *Malleus Maleficarum* (1486) by Heinrich Kramer: "In this profoundly misogynistic work, Kramer linked witchcraft entirely to what he regarded as women's spiritual weakness and their natural proclivity for evil. Above all, he linked witchcraft to supposedly uncontrolled female sexuality famously concluding that all witchcraft comes from the carnal lust, which in women is insatiable."

We wonder what Kramer would think of the decidedly nonlusty witches of modern cinema. In the children's horror film *Hocus Pocus* (1993), three witches are accidentally resurrected by unsuspecting children on Halloween night. Winnie (Bette Midler), Mary (Kathy Najimy), and Sarah (Sarah Jessica Parker) embody everything we believe about the prototypical witch. Along with their obsessed self-interest, they loathe children like the witches of fairy tales. The three witches' appearance, more specifically their supposed unattractiveness, is often pointed out for comedic effect. Like the green-faced Wicked Witch, they're destined to have visages that match their evil hearts.

Yet, is a witch's self-interest so terrible? Or is it an exaggeration of a male-driven society's fear of female rebellion? In *The Craft* (1996), Bonnie (Neve Campbell) uses her newfound witch powers to heal the painful and disfiguring scars on her back. Some might say this is vain, but wouldn't anyone channel their powers for similar reasons? It may be easier to judge Rochelle (Rachel True) when she finds revenge against her bully by casting a spell causing the girl to lose her hair, yet Rochelle is a teenager with a sudden windfall of supernatural abilities. She doesn't understand the gravity of the magic she is wielding. Unlike the cartoonish witches of *Hocus Pocus*, the women of *The Craft* are given more room to live in ambiguity. They are both good and bad, fully realized teen girls grappling with coming of age and the occult.

Eleven, too, would have been persecuted as a witch if she happened upon Salem, Massachusetts, a few hundred years ago. Although we think the Puritans would have been no match against her (or those pesky demogorgons)! We're just thankful that in the modern age, authentic female characters like Eleven can inspire a new generation of girls.

CHAPTER TWENTY

MISERY

Year of Release: 1990	
Director: Rob Reiner	
Writer: William Goldman	
Starring: James Caan, Kathy Bates	
Budget: $20 million	
Box Office: $61.3 million	

While the healer archetype exists in its purest form in horror, the usurpation of this trope is equally as prevalent. Perhaps no caretaker is more appropriately twisted as Annie Wilkes (Kathy Bates) in *Misery*. Based on the 1987 Stephen King novel, *Misery* is the story of infatuation gone violent in a gender reversal of more typical horror fare. Instead of a beautiful woman being stalked by a slasher hiding behind the folds of her shower curtain, author Paul Sheldon (James Caan) is bedbound, facing his "number one fan's" brutal allegiance. Because once Annie finds out that Paul has killed off her favorite character, Misery, from his romance novel series, she will do *anything* to alter that reality.

At first, it seems that Annie has saved Paul. She has bucked stereotypes and found the physical strength to pull Paul from the wreckage of his car accident. And like a true nurse subtype of the archetype, Annie tends to Paul's wounds with precision and care. He wakes to a pleasant, feminine bedroom with his own personal nurse. Considering the alternative of death, Paul seems to be lucky. He's escaped the brutal cold and is in a warm home with a woman who appears to have an inherent need to caretake. This is amplified by her actual profession as a nurse. But for every Nurse Nightingale, there seems to be a Nurse Ratched, and unluckily for Paul, Annie skews toward the latter. In his film review, Roger Ebert describes Annie's dichotomy: "Bates, who has the film's key role, is uncanny in her ability to switch, in an instant, from sweet solicitude to savage scorn."[1] This manifests in what becomes her immediate need to control and quite memorably crescendos with her gory hobbling of Paul's ankles with a sledgehammer. In one heart-pounding scene, as Paul struggles for a way to escape his captor as she's away, he comes across newspaper clippings of her past. It becomes clear that Annie has murdered patients at the hospitals where she's worked. She has warped her role: instead

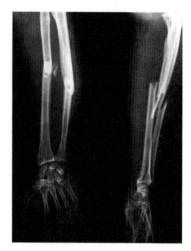

It takes an average of between six and twelve weeks for broken ankles or legs to heal.[2]

of caring for the vulnerable, she has exploited her power over them. This is a frequent theme in female-driven horror, often taking the form of deadly mothers. And, like mothers who kill their offspring, murderous nurses are also a stark reality.

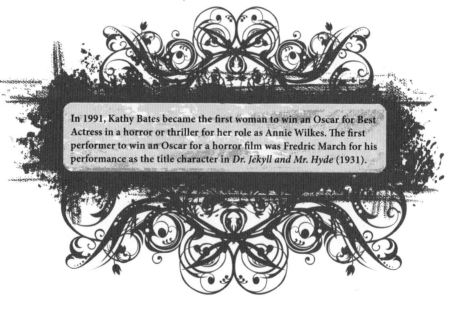

In 1991, Kathy Bates became the first woman to win an Oscar for Best Actress in a horror or thriller for her role as Annie Wilkes. The first performer to win an Oscar for a horror film was Fredric March for his performance as the title character in *Dr. Jekyll and Mr. Hyde* (1931).

"Jolly" or Jane Toppan was one of the most beloved nurses employed at Cambridge Hospital. She earned her nickname "Jolly Jane" from her pleasant and upbeat personality, and her friendliness toward her patients. By all counts, she was also one of the best nurses at the hospital, as well.[3] But in 1901, Jane Toppan's violence was finally exposed to the public. Not long after Lizzie Borden narrowly escaped a guilty verdict in Fall River, Massachusetts, Toppan was apprehended in the same state for murder. The nurse was believed to have killed thirty-one of her patients. Although, unlike Borden, Toppan didn't claim innocence. She, instead, informed the officials she was the cause of over a hundred murders and that her intention was to "have killed more people—helpless people—than any other man or woman who ever lived."[4] Toppan saw Cambridge Hospital as her own personal killing ground. Because of her medical knowledge, she was able to subtly administer drugs and fix charts to evade detection for decades. Another shocking aspect of the Toppan case is the supposed motive.

Women serial killers are rare a creature, but sexually motivated female killers are nearly unheard of. Yet, Jane Toppan, a woman born in the nineteenth century, explained that she derived an "erotic charge" from climbing into bed with her victims and holding their bodies as they died. One could argue that because of the time period in which Toppan operated, she, like Lizzie Borden, got off easy due to her gender. Because, despite her admission that she killed numerous patients and family members over the course of her life, Toppan was found not guilty by reason of insanity. Thus, she was commanded to a mental asylum for the rest of her days, rather than a prison.

Jane Toppan is by no means the only nurse to have been caught manipulating her role as caretaker. In a 2016 study published in the *Journal of Investigative Psychology and Offender Profiling*, researchers focused on the particular sort of killer profile that Jane Toppan and the fictional Annie Wilkes would fall under: "There is a small but growing body of evidence to suggest that HSKs (healthcare serial killers) tend to use their authority to target more vulnerable groups, such as the elderly, are more likely to be female than male, and, ironically given this finding, can be seen as confident men in how they go about their murders."[5]

It seems that the nurses killing their patients are able to maintain the face of the "jolly" nurse as described by Jane Toppan's patients, and as portrayed by Kathy Bates in *Misery*.

As described in Chapter Six (p. 40), another "caretaker" female archetype familiar to horror fans is the babysitter. Unlike a mother, the babysitter has less emotional investment in the children under her stead, but curiously, babysitters seem less likely to flip their scripts. They are frequently the victims of the antagonist, think clever final girls like Laurie Strode (Jamie Lee Curtis) in *Halloween* (1978) or long-suffering nannies like Greta Evans (Lauren Cohan) in *The Boy* (2016), who is tasked with caring for an inanimate doll. Or are they? Perhaps this fascination with the nonmaternal caretaker is a homage to the perpetuation of the urban legend centering on a young, vulnerable babysitter. This story, told at countless slumber parties, is outlined by Claire Linic at medium.com:

Jane was babysitting for the Millers for the first time. The Millers' kids, Bobby and Tiffany, were already asleep when Jane arrived. She was doing her homework at the kitchen table when the phone rang. "Hello?" Jane asked, but heard only heavy breathing in return. She hung up the phone and walked to the front door to make sure it was locked. The phone rang again. Jane picked up and said "Hello?"

"Have you checked the children?" said a low voice from the other end. Confused, Jane asked who was calling, but the caller was gone.

Fifteen minutes later, the phone rang again and the caller asked the same question: "Have you checked the children?" Jane knew she should go upstairs to check on Bobby and Tiffany, but her legs were too weak to safely climb the stairs. She called the Millers several times, but couldn't get ahold of them.

Next, Jane called the operator to see if it was one of her classmates trying to frighten her. The operator asked her to hold while she traced the calls. When the operator came back on the line, she told Jane: "Get out of the house now! I'll send the police. The calls are coming from inside the house!"[6]

When a Stranger Calls (1979) is a direct adaptation of this popular tale, and the film's iconic first scene was later referenced in *Scream* (1996) when Casey Becker (Drew Barrymore) becomes the franchise's first victim. In *When a Stranger Calls*, babysitter Jill Johnson (Carol Kane) fares better, though not before being stalked and traumatized by a murderous stranger. In fact, Jill is unable to fulfill her caretaker role properly, as the mysterious man upstairs kills the children under her watch. This is fascinating, as unlike a destructive mother trope, the babysitter is not the direct cause of the death of the children. All the same, because of "the man upstairs" played by Tony Beckley, Jill has ultimately failed as charge of the kids. Later, when the slasher escapes a mental asylum years later, he once again focuses his attention on Jill. A mother of young children, Jill goes on a date night with her husband, leaving the babysitter (Lenora May) to be terrorized. This mirroring, of Jill as mother and a younger girl as the babysitter, ultimately leads to a male-driven climax in which Det. John Clifford (Charles Durning) saves Jill, killing the proverbial bogeyman. This same sort of male intervention takes place in the aforementioned John Carpenter film *Halloween*, in which the babysitter Laurie fights valiantly against Michael Myers (Nick Castle), but it is Dr. Sam Loomis (Donald Pleasence) who shoots the slasher, ending the violence. Well, that is until *Halloween II* (1981), of course! As for Jill, she is resurrected in the 2006 *When a Stranger Calls* remake by Simon West. Played by Camilla Belle, Jill is allowed to be her own savior in this version, maiming her attacker so that he can be apprehended by law enforcement.

For many, babysitting is a rite of passage. It is a crossroad between childhood and adulthood, our first serious job. It, too, is inherently attached to the female, as "the history of babysitting is the history of the teenage girl."[7] As Miriam Forman-Brunell maintained in her book *Babysitter: An American History* (2011), "babysitting was seen as a compromise between teenage girls' desire for personal freedom and adults' expectations that they stay close to home."[8] Is there any wonder, then, why this tenuous state of womanhood is rife with horror? This led us to ask if there is any truth to the babysitter urban lore, and if so, why it has saturated popular culture so successfully.

Our first question led us to the murder of Janett Christman in March of 1950 as discussed in Chapter Six (p. 41). The thirteen-year-old eighth-grader at Jefferson Junior High School in Columbia, Missouri, skipped a school dance party in order to babysit for a local family. Just like in the urban legend that her story would inspire, when Christman arrived, her sole charge, a three-year-old boy named Greg Romack, was already asleep.

What happened next is the true, unyielding sort of horror that suspenseful movies are based on. Unfortunately for Janett Christman, she is unable to survive like the fictional Laurie Strode or Jill Johnson. Before she was brutally raped and strangled with a cord, Christman, known to be a talented piano player and an intelligent student with ambition, managed to get a call in to the local police. She urged them to "come quick," but the phone line was cut before they could trace the call or ascertain who she was.

When the Romacks returned home, sometime past 1:00 a.m., they were greeted by a horrific scene. Janett lay dead on the living room carpet, the cord still cinched around her neck:

> Inside of the home were clear indications Janett had resisted her attacker. Blood smears and fingerprints were found in the living room and kitchen, where the back door had been unlocked and left ajar. As the police followed the trail outside, the search dogs managed to track the assailant's scent one mile up from Stewart Road to West Boulevard and across West Ash Street before losing the trail. Back at the crime scene, an adult male's footprints were found near a side window of the residence that had been shattered with a garden hoe, where several authorities believed the perpetrator had gained entry, primarily due to muddy papers found on the piano that was situated nearby.[9]

One positive note: Greg Romack slept through the entire assault and was left unharmed. Yet, the perpetrator was never found, and nearly seventy years later, Janett Christman's murder remains unsolved.

So, why has this particular murder grown into the phenomenon of an urban legend? What seems to be the most integral factor is the emotion that it elicits. How did you feel after reading about Janett Christman's real murder? We know researching it brought up unpleasant and lasting horror within us. In their study, "Corpses, Maggots, Poodles and Rats: Emotional Selection Operating in Three Phases of Cultural Transmission of Urban Legends," researchers Kimmo Ericksson and Julie C. Coultas explain how these stories transcend time:

> The success in cultural evolution of stories such as urban legends, which are transmitted from individual to individual on a large scale, is to a considerable extent determined by the stories' ability to evoke widely shared emotions. These emotions may be both negative and positive. As a case study. Heath et al. (2001) focused on urban legends' ability to elicit disgust. Two main findings emerged. First, a study of web sites that document urban legends showed that legends with disgusting motifs were more likely to be spread across many websites. Second, in an experiment where urban legends were manipulated with respect to their capacity to evoke the emotion of disgust, participants expressed greater willingness in passing along those stories that were more disgusting (controlling for how other emotional and information factors were altered by the manipulation). The two findings together support the theory of emotional selection: A certain outcome of cultural evolution was demonstrated (viz., stories that evoke the emotion of disgust are prevalent among popular urban legends) which corresponded to an identified mechanism for emotional selection (viz., stories that evoke widely shared emotions are more likely to be passed along).[10]

The babysitter myth is rooted in painful reality, as is the "angel of mercy," a nurse who takes it upon herself to kill. In *Misery*, the nurse Annie Wilkes reflects back our expectations of a feminine caretaker, subverting what we have come to believe should be a selfless woman who waits on the vulnerable. It is these vulnerabilities of womanhood, whether we are a teenager tasked with too much responsibility or a nurse given too much power, that create authentic horror.

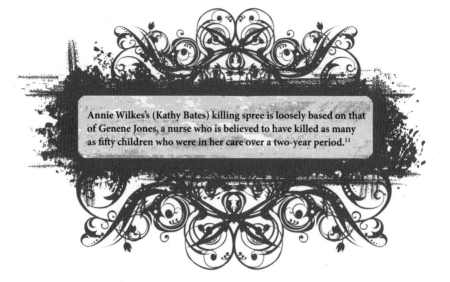

Annie Wilkes's (Kathy Bates) killing spree is loosely based on that of Genene Jones, a nurse who is believed to have killed as many as fifty children who were in her care over a two-year period.[11]

CHAPTER TWENTY-ONE
NAILS

Year of Release: 2017	
Director: Dennis Bartok	
Writer: Tom Abrams, Dennis Bartok	
Starring: Shauna Macdonald, Steve Wall	
Budget: unknown	
Box Office: $41,000	

In the 2017 Irish horror film *Nails*, the main character, Dana Milgrom (Shauna Macdonald), is struggling through the most challenging time of her life. As horror author Joe Hill once said, "Horror was rooted in sympathy. In understanding what it would be like to suffer the worst." This resonates in Dana's story, as she has recently been paralyzed by an accident and is unable to speak or move. This is a particularly cruel fate, as we come to learn that Dana's life centers on her ability to be physical. She is an athletic coach, devoted to yoga and jogging, who has had her livelihood ripped away. As the movie unravels, we see she also is going through a trying time with her husband and teenage daughter, all while being confronted by a malevolent ghost in her hospital room. To stay true to the Female Gothic tradition, Dana's husband, Steve (Steve Wall), does not believe in her claims of a spirit lurking in her room, and as we are seeing in a distinct pattern, he attributes what she deems as fact as emotions bubbling to the surface in response to the recent trauma of her accident.

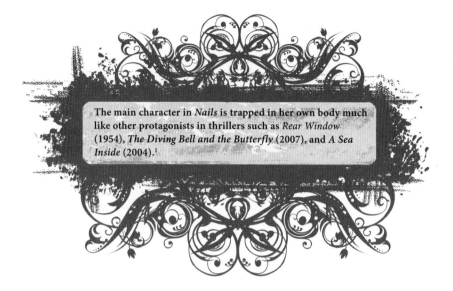

The main character in *Nails* is trapped in her own body much like other protagonists in thrillers such as *Rear Window* (1954), *The Diving Bell and the Butterfly* (2007), and *A Sea Inside* (2004).[1]

Dana perfectly exemplifies what is a common female trope in horror media known as the wounded healer. Many horror heroines are injured through the course of their ordeal, but Dana's paralysis is more than a surface trauma. She begins her journey this way and is already tasked with the grief of losing her body to the accident. A body that she used for both work and hobby and that is indicated to be an important quality to her husband. This grief of her former life, of her inability to mother her daughter, to communicate, to be who she formerly was, is a vital piece of the wounded healer. This female character must have a severe trauma in her past that will assist her in the ability to survive her current horror scenario. Author Stacey L.L. Couch explains the archetype:

A person with the wounded healer archetype is able to draw on the experience of her own suffering to generate boundless compassion. In the shadow, this manifests as the bleeding heart that gives well beyond her capacity to people with similar woundings. In the light, this is the battle-hardened champion that is able to show up in a relevant and particular way that provides just the right remedy at the right time. The wounded healer archetype implies an intimate understanding of agony. Sometimes the wounded healer is able to heal her own wounds and use this knowledge to help heal others.[2]

If we reflect back on Dana's experience in *Nails*, she is still navigating through the physical trauma of her accident but has garnered an ability to communicate with the terrifying specter in her presence. This ultimately leads to her discovery that she has a connection with him, a connection that she chooses to exploit in order to save the proverbial day. Other examples of the wounded healer archetype exist in numerous horror films. In both versions of *It* (1990, 2017), Beverly Marsh (Emily Perkins, Sophia Lillis) comes from a traumatic, troubled home. She is beaten almost daily by her father. This instills in her a cleverness and a need to overcome and, in theory, to find the strength to defeat her father. This is why it is Beverly who must inflict the final blow to the menacing Pennywise. In *Hush* (2016), Maddie Young (Kate Siegel) also has a physical "wounding" like Dana in *Nails*, and she uses her deafness as an advantage, ultimately defeating her

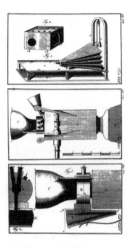

An example of a speech synthesizer by Wolfgang von Kempelen from 1791. The main character in *Nails* uses a modern speech synthesis machine to communicate.

attacker by relying on what makes her both real and flawed. Eleanor, in the aforementioned *The Haunting of Hill House*, is also a wounded healer, as her past with a controlling mother seems to allow her a heightened level of empathy and, in *The Haunting*, the ability to sacrifice herself for the greater good.

There is often a healer archetype in religious horror films, particularly with nuns. They are often depicted as evil in contrast of what nuns are supposed to be like. We asked the writer and director of *The Devil's Doorway* (2018), Aislinn Clarke, why she thinks that is:

Being a nun is a pretty extreme stance: chastity, piety, isolation, devotion, poverty, servitude. It is quite something to live up to. Growing up as a girl in Catholicism, becoming a nun is the dream for many girls: it is the closest thing to being the Virgin Mary and the Virgin Mary is beloved by everyone. If you have to be a woman, she's the woman to be. But you can't be her. Thus, nuns are a symbol of a certain female ideal and the guaranteed failure to meet it. They are women who are pure, unsexed, unattainable to men; there is a pleasure in their corruption and, as the bar is set so high, they are corrupted just by being shown like other women, with a woman's desires and a woman's faults. Still, horror films tend to that corruption to extremes; it's a mockery of the ideals, a spotlight on hypocrisy, or a perverted kick from seeing it all turned upside down. On screen, the nun archetype can be inverted as easily as a cross. However, for some of us, it is also plays on a genuine fear that goodness is a veneer and that institutions are riddled with duplicity, and that no one can be trusted. Of course, anyone who has ever been taught by nuns or been around them knows that they are just like other women. They can be cruel, they can misuse their power, they can give in to their lusts and their caprices. No archetypal figure can ever manifest as how they "are supposed to be" in real life. In my film, I wanted the nuns to be individuals. Although many of them are evil or did evil things—as plenty of nuns in the Magdalene Laundries were and did—there are also those that try to live up to their vows. They are all individuals making choices in a corrupt theocratic system. They are all women who have taken on the archetypal role of nun: some play to type, some play against, but there are no perfect nuns, because there are no perfect people.

The Descent (2005) is another example of the wounded healer archetype, as the "final girl" Sarah (Shauna Macdonald) is grappling with the death of her husband and child. This grief has altered who she is and, in the end, gives her a strength and resilience that allow her to be the last woman standing after a girls' spelunking trip gone wrong. We can't help but root for the wounded healer. She is often stronger for her painful past, an aspect of humanity we can all relate to. Dana in *Nails* ups the healer ante at the end of the film by sacrificing her life in order to save her daughter. What could encompass the healer more than that? This final act of complete selflessness makes a hero healer. These heroes are the women who don't make it to the last frames of the movie. They are not destined to be the final girls, but their sacrifice, like that of the archetypal mother, is always appreciated. In *Tales from the Crypt: Demon Knight* (1995), Irene (C.C.H. Pounder), who has been bitten by a demon and has little time to live, bravely ushers the protagonist, Jeryline (Jada Pinkett Smith), to safety before going out in a literal blaze of glory. Another hero healer would be Evelyn (Alfre Woodard) of *Annabelle* (2014). Evelyn, like Sarah, is also a wounded healer because of her struggle with her daughter's death. At *Annabelle*'s climax, Evelyn fulfills her archetype of a hero healer by not allowing Mia (Annabelle Wallis) to let the nefarious entity inside the doll take her soul. Instead, Evelyn grabs Annabelle and plummets from the window, saving everyone. One aspect of this hero healer trope that must be noted is that it's overwhelmingly portrayed by people of color as described in "22 'Heroic Deaths' by Black Characters in Horror Movies": "We have the phenomenon of black 'heroic deaths' in which black characters (usually peripheral) voluntarily sacrifice themselves—or at least, volunteer for tasks that mean certain death—in order to save the (usually white) main stars."[3] See Chapter Five (p. 33) for more exploration on race in horror.

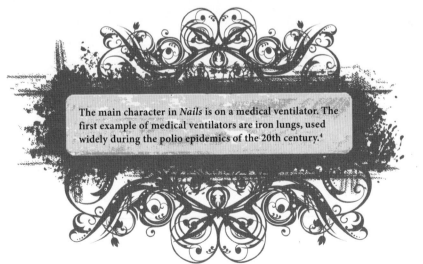

The main character in *Nails* is on a medical ventilator. The first example of medical ventilators are iron lungs, used widely during the polio epidemics of the 20th century.[4]

Heroes come in many forms. Whether it be the hero healer like Eleven in *Stranger Things* or the subverted caretaker, we hope the next time you have a hospital stay you find yourself healed by the former rather than the latter!

SECTION EIGHT
HYSTERIA

THE HAUNTING OF HILL HOUSE

Year of Release: 2018	
Created by: Mike Flanagan	
Starring: Carla Gugino, Kate Siegel	
Network: Netflix	

"Don't do it, Eleanor told the little girl; insist on your cup of stars; once they have trapped you into being like everyone else you will never see your cup of stars again; don't do it; and the little girl glanced at her, and smiled a little subtle, dimpling, wholly comprehending smile, and shook her head stubbornly at the glass. Brave girl, Eleanor thought; wise, brave girl."

—Eleanor Vance, *The Haunting of Hill House*

In 1959—the same year John Knowles's World War II classic *A Separate Peace*, Robert Bloch's *Psycho*, Robert A. Heinlein's *Starship Troopers*, and Richard Condon's *The Manchurian Candidate* were published—a woman's name appeared in the bestseller lists of the world. Already a prolific short story writer, Shirley Jackson was best known for her rather polarizing story "The Lottery" published in 1948 by *The New Yorker*. As noted at ShirleyJackson.org, "The Lottery," about a brutal ritual performed by the denizens of a rural town in New England, "generated the largest volume of mail ever received by the magazine—before or since—almost all of it hateful."[1]

While this swell of complaints may seem to indicate that "The Lottery" must have died in obscurity, the opposite occurred, as it is one of the most well-known short stories of the modern age. We would even presume to guess most readers of this book were assigned "The Lottery" in school!

Eleven years later, a time in which literature was still dominated by men, Jackson's novel *The Haunting of Hill House* arrived to immediate attention and approval. As described in a more recent piece for the *New York Times*, "Shirley Jackson's the *Haunting of Hill House* is considered not only one of the finest horror novels of the 20th century, but also the definitive haunted house story. It was a finalist for the National Book Award upon its publication in 1959."[2]

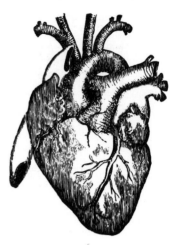

Author Shirley Jackson was only forty-eight years old when she died of cardiac arrest due to a coronary occlusion. This is believed to be the same manner Tsar Nicholas II died in 1917, aged fifty.

It is possible to die of fright. Being literally scared to death resembles a heart attack as a sudden burst of adrenaline hits your heart.[3]

On the surface, *The Haunting of Hill House* seems to be a straightforward ghost story. It has all the familiar trappings of what we have come to expect in the genre: a spooky, old house with a rich and tragic past; strange, subtle happenings that haunt the unsuspecting residents. Yet, Jackson's genius lies in her ability to simultaneously entertain while exploring the depths of womanhood. This merging of the gothic and the female has a long and famous past. Charlotte Brontë brought us *Jane Eyre* a hundred years before *The Haunting of Hill House*. Both books involve a seemingly weak heroine trapped in an imposing house, haunted by a past she can't alter.

Only a few years later, the first adaptation of Jackson's novel hit the big screen. The 1963 version *The Haunting* stayed faithfully close to the source material, balancing both the supernatural with Eleanor's psychological breakdown. Directed by Robert Wise, also the director of *The Curse of the Cat People* (1944) and *The Body Snatcher* (1945), *The Haunting* provided decent scares, cementing itself as one of the first popular haunted house films. In the book *Robert Wise on His Films* (1995), the director explained to author Sergio Leeman why he accepted the challenge of bringing Shirley Jackson's novel to the screen: "that he was attracted to the project because the book made the hair curl on the back of his neck."[4]

Forty years after *The Haunting of Hill House* emerged, a reimagined film was released by DreamWorks Pictures. *The Haunting* (1999) boasted both a big budget and an all-star cast, including Catherine Zeta-Jones (Theodora) and Liam Neeson (Dr. Marrow). This version steered quite far from Jackson's novel. While Eleanor, played by Lili Taylor, shares similarities with her literary counterpart, and Hill House is equally as menacing, the plot is drastically changed. This, coupled with the film's heavy reliance on CGI, may have been the reason for its demise. *The Haunting* was panned by critics and nominated for five Razzies, awards devoted to recognizing the worst films of the past year.

But, like Hill House itself, *The Haunting of Hill House* lives on. In 2018, Mike Flanagan, director and writer of popular female-driven films such as *Gerald's Game* (2017) and *Hush* (2016), took on the incredibly complex task of adapting Jackson's novel into a television series. Although there are quite a few changes from the novel—for instance, the characters of Luke, Eleanor, and Theodora are now siblings in the television series—Flanagan captured the heart of Jackson's work. It is the complicated ties of relationships and their warring psyches that propel *The Haunting of Hill House* forward into a satisfyingly scary exercise in grief.

In the 2018 series, the family matriarch Olivia Crain (Carla Gugino) is slowly driven mad within the walls of Hill House. This madness eventually leads her to be abandoned by her husband and children as they flee for their lives. What is significant is that Olivia, in her death, becomes part of the house. She is as permanent, if not more so, as the structure itself. This fusion of mother and home is a theme explored by Shirley Jackson in her novel and other works, as outlined in the 1996 study "House Mothers and Haunted Daughters: Shirley Jackson and the Female Gothic" by Roberta Rubenstein of American University. Rubenstein explains the torment of separating oneself from her mother: "The tensions between mother/self and between home/lost connote a young child's ambivalent desires and fears: both to remain merged with the mother (who becomes emotionally identified with home) and to separate from her, with the attendant danger of being lost."[5]

This conundrum is on full display in Flanagan's TV series, as the adult children of Olivia, particularly Nell (Victoria Pedretti), grapple with the desire to be with their mother, therefore to succumb to the pull of Hill House, or to fully disengage from her and the structure itself. In Jackson's novel, the character of Eleanor, or Nell, has lived a reclusive life, taking care of her disabled and demanding mother. While her relationship with her mother, Olivia, is different in the more recent adaptation, there still lies that universal tension that is a trope of the female gothic tradition. As Rubenstein explains:

> In these narratives authored by women and focusing on female protagonists, traditional elements of the Gothic genre are elaborated in particular ways. Notably through the central character's troubled identification with her good/bad/dead/mother whom she ambivalently seeks to kill/merge with; and her imprisonment in a house that, mirroring her disturbed imaginings expresses her ambivalent experience of entrapment and longing for protection."[6]

In both the novel and the 2018 series, Eleanor's despair ends in her death on the grounds of Hill House. And in both instances, it is left to the audience to decide if she is a disturbed woman who committed suicide or a victim at the hands of a malicious, supernatural entity. If her death

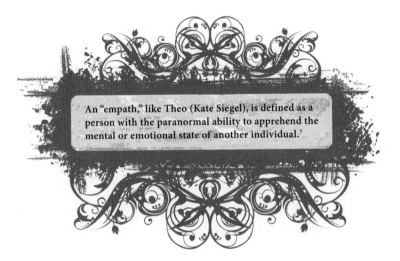

is her own doing, based on Rubenstein's observations, this may reflect her unconscious desire to merge with her mother, perhaps more blatantly in Flanagan's adaptation, as Nell "comes home" to where she not only lived with her mother, but where her mother remains.

An "empath," like Theo (Kate Siegel), is defined as a person with the paranormal ability to apprehend the mental or emotional state of another individual.[7]

This mental erosion and its inherent link with the supernatural is an integral aspect of the gothic, particularly the female variety as outlined in "Ghosts of the Mind: The Supernatural and Madness in Victorian Gothic Literature":

When reading and reviewing Gothic texts, the prevalence of ghosts, mysterious apparitions, and unexplainable sounds and events is apparent. Just as frequent, however, is the theme of insanity—of hallucinations, anxiety, and complete mental break-down—particularly in Gothic texts' weakest female characters. Although the occurrences of insanity and the supernatural may seem coincidental or unrelated, a closer examination of the culture surrounding such literature tells a different story.[8]

At a time when medicine and the understanding of psychology was embryonic, authors of gothic literature in the Victorian era sought explanation of mental illness through the supernatural. A well-known example of this cohesion of madness and the macabre is Henry James's

The Turn of the Screw (1898), in which the unnamed governess is in mental anguish over the presence of ghosts. Like many of her fictional counterparts, the governess is told that what she is seeing is not real, and because of her place in society, as both a female and of a lower class, she is inevitably driven mad with the severe chasm between what she is told and what she believes. This is on display in *Jane Eyre*, which has been adapted into dozens of films, plays, and even manga. Jane is in the same hierarchical disadvantage as the governess and thus is unable to investigate the strange goings-on in Thornfield Hall. She is at the mercy of Edward's explanation, as men, particularly men of a higher class, were considered the bearers of truth. The irony of *Jane Eyre*, of course, is that Edward's truth, that his mentally ill wife is chained in the attic, will lead to his second love's similar madness if he doesn't allow Jane to understand reality. While *The Haunting of Hill House* can hardly be considered Victorian, and it was adapted to television in the modern age, it still contains this trademark tug-of-war between female experience versus male explanation.

In Flanagan's series, the Crain children, both female and male, grow up believing a warped version of the truth. Hugh Crain (Henry Thomas/ Timothy Hutton), the patriarch, takes it upon himself to direct the narrative of their last night in Hill House away from reality. He feels it is his responsibility to shelter those in his family he deems weak. This includes all three of his daughters, as well as Luke (Oliver Jackson-Cohen), the youngest son who is portrayed with effeminate traits and, later, as a drug addict. This is in stark contrast to his treatment of his eldest son, Steven (Michiel Huisman), who shares in more truth than his siblings. This, interestingly, creates in Steven an aversion to the supernatural and, most notably, the home, a naturally feminine sphere of influence. This is depicted in how he both holds the most resentment against his mother and Hill House and his current predicament as a man thrown out from his domestic home by his wife. The latter manifests because Steven chooses to emulate his father. In the same way that Edward Rochester lied about his mad wife in the attic, and Hugh Crain lied to his children about the nature of their mother's demise, Steven feels it is his masculine duty to keep the unpleasant truth from his wife. While she unknowingly pines for a child, frequenting a fertility clinic, Steven keeps his past vasectomy a secret. While he inwardly justifies this cruelty, Leigh (Samantha Sloyan) is left in the metaphoric dark, unable to become a mother while blaming her own body.

We now know this treatment by Steven, a learned behavior from Hugh, is a form of manipulation and abuse termed gaslighting: "Gaslighting is a tactic in which a person or entity, in order to gain more power, makes a victim question their reality."[9] Nearly all the brooding heroes of Victorian gothic literature used this technique to varying degrees. Sarah Zettel further explains in her article "Are We All Gas Lighters? How Crime Fiction Helps Us Understand the Part Communities Play in Continuing Abuse" that "he can't fine tune the tale and his presentation. He appears entirely reasonable. That's because he's the one who knows what's really going on. On the other hand, when the victim tells her story to the outside world, she might well look disheveled, frightened, doubtful, angry, shrill, hysterical."[10]

It is this presumed hysteria, a fire stoked by male influence, that is prevalent in horror from its literary inception. One more modern example of this dynamic is the Robert Zemeckis film *What Lies Beneath* (2000). The suspenseful movie, a box-office smash, embodies many aspects of the female gothic. When Claire Spencer (Michelle Pfeiffer) begins to sense a ghostly presence in her home, her husband, Dr. Norman Spencer (Harrison Ford), publicly makes fun of her. As a man of science, he is quick to point out any bit of emotion Claire displays, often attributing her panic or distress to past trauma. He continually ignores, berates, and dissuades her from believing what she has witnessed with her own eyes. This poisonous interplay explodes in violence, as Norman's true self is revealed. Claire's intuition has been correct all along: her home, her feminine sphere of influence, has been invaded. Her husband has murdered a woman, one of intellect and promise, all because he was afraid for his reputation. In the end, the audience is aware that Claire's "hysteria," like that of so many female gothic protagonists before her, is justified emotion.

These sociological complexities, most vitally a strain between the male and female, have been evident in horror since Mary Shelley's *Frankenstein*

(1818). And too, there is a natural element of fear in the female's role as keepers of home, often ones with shadowy corridors and ghosts perched, ready to frighten. Homes that, much like society's stifling rules, swallow women whole. As Eleanor describes in *The Haunting of Hill House*, "I am disappearing inch by inch into this house. I am going apart a little bit at a time because all this noise is breaking me."

CHAPTER TWENTY-THREE
HEREDITARY

Year of Release: 2018	
Director: Ari Aster	
Writer: Ari Aster	
Starring: Toni Collette, Milly Shapiro	
Budget: $9 million	
Box Office: $79.4 million	

In the Female Gothic tradition, Charlotte Perkins Gilman wrote her enduring masterpiece "The Yellow Wallpaper." Published in 1892, Gilman's short story contains one of the most disturbing and memorable endings ever printed. Like her contemporaries, Gilman pushed the boundaries of literature, choosing to give voice to common female struggles through the art of words. As Greg Johnson detailed in his piece "Gilman's Gothic Allegory: Rage and Redemption in The Yellow Wallpaper," Gilman "adroitly and at times parodically employs Gothic conventions to present an allegory of literary imagination unbinding the social, domestic, and psychological confinements of a nineteenth-century woman writer."[1] In "The Yellow Wallpaper," the main character is nameless (a familiar device used in other Female Gothic works like Daphne du Maurier's *Rebecca* [1938] and the aforementioned *The Turn of The Screw*) and speaks to the reader through diary entries. What is significant about the narrative of "The Yellow Wallpaper" is its gothic telling of women's reality. At a time when the term "postpartum depression" was in no one's vocabulary, Gilman's story examined both its effect on the new mother and, more important, how she was viewed by those around her. Instead of finding empathy and understanding, the nameless narrator is locked away. She is driven far madder by her boring, lackluster surroundings than by her initial depression. This, hauntingly, leads her to find unnatural interest in the wallpaper of her room.

Hereditary, a horror film that was released in 2018 to much acclaim, tackles some of the same Female Gothic notions as "The Yellow Wallpaper." As the name suggests, motherhood is at the forefront of *Hereditary*. Annie Graham (Toni Collette) is processing her grief and frustration over the death of her own distant, difficult mother. Much like the characters populating Shirley Jackson's work, Annie's relationship with her mother is central to who she is, and who she wants to be. She has spent her life striving to be different, to be a more present mother. The irony, which the audience sees in a dream sequence, is that Annie had considered aborting her first-born son. Although she could never admit her trepidations about motherhood freely, there is a deep place in her psyche where she can express her fears. While *Hereditary* takes place over a hundred years after "The Yellow Wallpaper," it exemplifies the same reality: women are still not allowed to admit their darkest and basest fears and desires. The disconnect between Annie's inner and outer lives is alluded to again, when it is discovered that while sleepwalking in the past, she nearly set her son Peter (Alex Wolff) on fire. This unconscious desire to cause destruction of the male is eerily similar to Gilman's story.

Then, while navigating her own role as daughter, Annie is met with the shocking death of her preteen daughter. Done with slow and unsettling effect, Annie finds the decapitated head of her daughter, Charlie (Milly Shapiro), who is a victim of an accident. An accident, notably, caused by the very son whom Annie had had an unconscious desire to eradicate. What comes next is a raw look inside a mother's grief. Annie wails, screams,

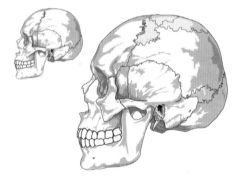

Milly Shapiro, who portrays Charlie, has cleidocranial dysplasia (CCD), a condition that affects the bones and teeth. Gaten Matarazzo (Dustin in *Stranger Things*) was also born with CDC, which has found new attention thanks to these young stars![2]

and bargains with the universe to let her die. This sort of emotionally charged grief is often tied to the feminine and exploited or made fun of in horror media, as men are deemed the voice of logic, while women become hysterical, allowing their emotions to run amok. This leads us to the notion of "female hysteria," an antiquated diagnosis.

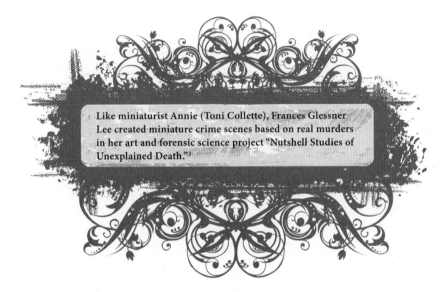

Like miniaturist Annie (Toni Collette), Frances Glessner Lee created miniature crime scenes based on real murders in her art and forensic science project "Nutshell Studies of Unexplained Death."[3]

For hundreds of years, female hysteria was a catchall to describe any action by a woman deemed inappropriate or unusual. Even women with hearty sex drives were labeled hysterics by medical doctors: "In the Victorian era most women carried a bottle of smelling salts in their handbag: they were inclined to swoon when their emotions were aroused, and it was believed, that, as postulated by Hippocrates, the wandering womb disliked the pungent odor and would return to its place, allowing the woman to recover her consciousness."[4]

Other common symptoms attached to female hysteria were "faintness, nervousness, insomnia, fluid retention, heaviness in abdomen, muscle spasm, shortness of breath, irritability, loss of appetite for food or sex, and 'a tendency to cause trouble.'"[5] This last symptom is the most significant as a symbol of the feminist movement, for if women in the Victorian era dared to live alternatively to their upbringing, they were seen as sufferers of mental disease. This reached a fever pitch at the time of Gilman's writing, as women were being shuttled to asylums for the slightest of behavioral differences. Journalist Wendy Wallace of the *Daily Mail* came upon dozens of bleak, unsmiling portraits of the many women locked up in England's Bethlem (also known as Bedlam) Hospital during the Victorian era. There was Emma Riches, whose "insanity was caused by childbirth"; Sarah Gardner, deemed "insane because of anxiety"; and, as Wallace recounts:

> Eliza Josolyne, twenty-three, was admitted to Bethlem in February 1857, with the cause of her apparent insanity recorded as "overwork." She looks distraught and her face bears marks of injury. Eliza had been the only servant in a twenty room house and was unable to keep up with the work over the hard winter months when every room would have required a fire burning in its grate and lamps to be lit early. Would doctors now diagnose burnout and acute stress?[6]

While some of Annie's stresses in *Hereditary* are more modern, like an impending art show, her grief and maternal guilt are timeless. So, when she begins to unravel a supernatural mystery surrounding her dead mother and daughter, Annie's husband Steve (Gabriel Byrne) believes Annie is suffering from hysteria. This is exactly how the husband responds in "The Yellow Wallpaper" and has become a popular trope in horror media. A female character is often faced with unbelievable truths, and thus, the males in her life label her emotions and conclusions as hysterical, especially when she has recently gone through a time of grief or hardship. This is on full display in movies like *Scream* (1996), when Sidney Prescott (Neve Campbell) is seen as weak and hysterical because her mother was previously a victim of murder, or in the case of Charlotte Hollis (Bette Davis) in *Hush, Hush Sweet Charlotte* (1964). Because her lover was murdered and she was considered a suspect, Charlotte spends her life in seclusion, a constant topic of the town gossips. Because she is considered the neighborhood eccentric, children run in fear from Charlotte's house, all because she was in close proximity to a murder. This murder has defined her life, creating an overwhelming grief that she never escaped. Her emotional state, or one could argue "female hysteria," is then exploited by her loved ones in order to destroy her. Although, in true horror fashion, Charlotte discovers their schemes and destroys them first. This victory, and the realization that Charlotte is not to blame for the murder, give her a sort of peace that she hadn't been afforded until old age.

King Paimon, a demon who takes physical form in *Hereditary*, first appeared in the 1600s in a tome called *Lesser Key of Solomon.*[7]

Annie, like Sidney and Charlotte, is right about her suspicions. There are devils at work, infiltrating her female sphere of home, and they are eventually successful in killing both the males and females of the Graham family. Although it is integral to the themes of *Hereditary* to note that Charlie lives, sloughing off her female body for what is considered a better, male one. She is able to do what her mother was not: eradicate the male and, one better, usurp his dominance.

CHAPTER TWENTY-FOUR
THE OTHERS

Year of Release: 2001	
Director: Alejandro Amenàbar	
Writer: Alejandro Amenàbar	
Starring: Nicole Kidman, Fionnula Flanagan	
Budget: $17 million	
Box Office: $209.9 million	

While grief and postpartum depression are perfect fodder for the Female Gothic tradition, there are further universal struggles of womanhood that have been around since the advent of time. War, on its surface, is a man's onus. Millions of men have died on the battlefield; whether drafted or there of their own accord, they fought for their country, their government, their freedom. Yet, war extends farther than the bloody fields where battles are won and lost. It reaches its ruining, toxic fingers into the lives of women, too. This extends to women all around the world, as described by Katherine R. Jolluck of Stanford University in "Women in the Crosshairs: Violence Against Women during the Second World War":

Though sexual violence—outside of the brothel system—was neither ordered nor officially condoned, it was nevertheless ubiquitous in Nazi-occupied Poland and western USSR. The topic of sexual abuse of males remains understudied, but it is clear that German officials and soldiers at all levels used sexual violence against females to assert power and control, to humiliate the occupied population, and to act out violent racist and misogynistic feelings.[1]

Jolluck continues, explaining how far this sexual abuse reached. "It was not just the German occupiers who raped; women suffered sexual violence from local men of different nationalities and also from conationals. In four specific settings females—particularly but not exclusively Jews—were highly vulnerable to sexual abuse: in hiding, ghettos, camps, and partisan units."

While women of all nationalities were raped and murdered during World War II, some women were also left behind. They were widowed, left with no money or prospects, at a time when society still deemed women as the weaker sex. In stark contrast to the glossy posters depicting female volunteers, many wives and mothers were left homeless with broods of children. In *The Others*, we see this devastation firsthand, as Grace Stewart (Nicole Kidman) lives alone with her children, clearly widowed by the Second World War, though she's not keen to admit it. Grace is one of a vast many women left to pick up the pieces of a war started, and fought, by men.

Writer and director Alejandro Amenàbar uses a familiar gothic aesthetic to showcase the misery and loneliness of this postwar existence. Grace and her children Anne (Alakina Mann) and Nicholas (James Bentley) live in rural England, far away from the bomb damage of London. Their manor could be plucked from any Victorian novel, obscured by fog and creaky in every way. Even the electricity is spotty because of the just-ended war.

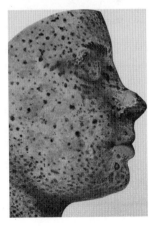

If Grace's seclusion is not difficult enough, she must contend with an inherited disease suffered by both of her small children. Anne and Nicholas have a rare disease that makes them unable to be in sunlight. This, again, makes for more gothic imagery, as the house must be dark at all times. Every curtain is drawn, creating an intense inner world of mother and children that is unhealthy.

An example of a child with Xeroderma Pigmentosum in 1906, the same disease the children have in *The Others*.

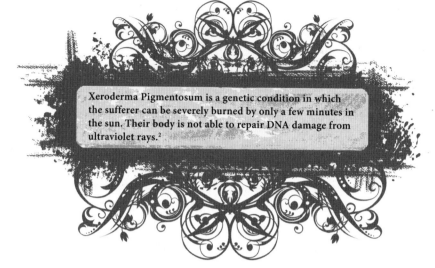

Xeroderma Pigmentosum is a genetic condition in which the sufferer can be severely burned by only a few minutes in the sun. Their body is not able to repair DNA damage from ultraviolet rays.[2]

This forced isolation and reduced social interaction bring back shades of "The Yellow Wallpaper." As the film continues, we start to recognize a fracturing within Grace. Her emotions are exaggerated, oscillating between panic and sternness. She has so much on her shoulders, including strange goings-on in her old house. And when her husband mysteriously comes back only to leave her again, it becomes clear to the audience that Grace's mental well-being is at risk. Like with Annie in *Hereditary*, the audience is left to wonder if Grace is creating these ghost happenings in her mind. Are these supposed "others" real? Or, is this a woman on the cusp of a mental breakdown?

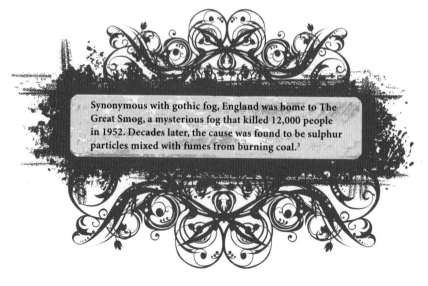

Synonymous with gothic fog, England was home to The Great Smog, a mysterious fog that killed 12,000 people in 1952. Decades later, the cause was found to be sulphur particles mixed with fumes from burning coal.[3]

The Others has perhaps one of the best twist endings in recent ghost film memory: we learn that Grace and her children are indeed "the others." They are the ghosts haunting this gothic mansion. All because Grace *did* lose her mental faculties. After her husband's death in the war, and being left alone with children who are unable to go outside, Grace makes a dark and complex choice. As with the narrator of "The Yellow Wallpaper," this isolation and repression causes her to lash out. She smothers her children in their sleep. Again, this brings back the themes of *Hereditary*, of a mother's deep and unspoken desire to kill her children. As we all know from the press, this is a real phenomenon that can occur when women's mental illnesses are left unchecked. Andrea Yates is one real example of a woman who had outwardly struggled with postpartum psychosis and schizophrenia before she drowned her five children in a bathtub in 2001. Because of her well-documented ailments, Yates was found not guilty by reason of insanity and thus committed to a Texas state mental hospital. It is no wonder, in the centuries before Andrea Yates was born, that women who had less chance of help or scientific understanding resorted to such extreme actions. In *Madness is a Woman: Constance Kent and Victorian Constructions of Female Insanity*, Samantha Pegg outlines similar cases in the Victorian era:

The 1860s saw a substantial number of cases in which mothers murdered their children and raised an insanity plea. Anne Brough cut the throats of her six children while, as she stated, "there was something of a black cloud over my eyes." *The Times* were so aggrieved by the jury finding of insanity that they saw fit to launch an attack upon the verdict and the "vindictive woman." Anne had, as *The Times* pointed out, at least one affair, probably more, and her husband had separated from her, leaving her alone in the family home with the children. Her confession was detailed (a fact *The Times* rightfully stated should have destroyed her insanity plea) and the best evidence the medical authority in the case could offer was that Anne suffered from "blood to the head" and strong and uncontrollable homicidal impulse.[4]

Andrea Yates, too, over a hundred years later, was treated similarly by the American press.

Another important facet of the Female Gothic tradition is the ghost. The archetypal picture of a ghost seems to be inherently female. Ghosts often wear long, flowing clothes and are beautiful, ephemeral young women. These ghosts can often be women whose presence permeates the feminine sphere of home, as in the memory of Rebecca in *Rebecca*, the very alive Bertha Mason in *Jane Eyre*, or the more classic example of Olivia Crain in Flanagan's *The Haunting of Hill House*. Like Olivia, who haunts Hill House, many legends, believed to be real, center around feminine spirits: "The White Lady," "Bloody Mary," and "La Llorona" are several examples. Many of these ghost myths are about women scorned. Asian filmmakers have found success by exploring the female ghost myth in recent years, capitalizing on several horror tropes, including revenge like in *Ringu* (1998) and *Ju-On: The Grudge* (2002). Usurping our expectations of a pleasant female ghost in a white aura, the ghosts of these films use their feminine traits (long black hair) to terrify.

In a pursuit to better understand the Female Gothic tradition and how it translates on film, we interviewed experimental filmmaker and film professor at DePaul University Shayna Connelly.

Meg: **"You've described yourself as a fan of ghost stories from a young age. What were some gothic tales that caught your imagination, and can we recognize their influence in your films?"**

Shayna Connelly: "I was caught up in the line between fiction and nonfiction early in regard to ghosts, and that was more critical to my obsession than Poe, Lovecraft, or Jackson were. When I was six, I discovered a book on haunted houses in the nonfiction section of my library. It was a kid's book with photos of ghosts and the stories behind the hauntings. Among the most convincing photos was 'The Brown Lady,' a ghost in England who appeared as a classic hazy person-shape gliding down the stairs. Discovering this book in the nonfiction section framed ghosts as part of the natural rather than supernatural world and set up a conundrum for me about the line between truth and the unknown."

Meg: "I can relate! I was drawn to similar books at my local library, the creepier the photos the better!"

Shayna Connelly: "Many years later ghosts kept cropping up in my work, but ghosts don't scare me. People do. That means my cinematic ghosts are not the terrifying kind. They function more like an enlightened being or superior alien; think 'ghost as philosopher,' rather than 'ghost as terrifying enigma.' I'm fascinated by knowledge that is beyond our grasp and envision a ghost as a being that serves as a bridge to understanding and truth. What I love in terms of ghost stories are those told to me by people who've experienced them; so, folklore and personal narrative drive me. I don't read much gothic fiction and prefer true crime, psychological thrillers (Ruth Rendell/Barbara Vine, Patricia Highsmith, Tana French), memoir, and nonfiction books about death. These written genres definitely influence my work. That said, I recently reread Shirley Jackson's *The Haunting of Hill House* and appreciated her writing, as well as how the book revolutionized the haunted house. Prior to Jackson, ghosts rattled around inside buildings, which meant they could be chased out. She created the malevolent house where the spirit fused with the architecture, making the structure itself evil. This paved the way for *House of Leaves* (2000), a sprawling haunted house story/double narrative that I find much more compelling than fiction written prior to Jackson. *House of Leaves* is also experimental in form and concept, which is another big influence on me as a filmmaker."

Kelly: **"The theme of haunting has been explored in depth in your work. Why do you think ghosts and haunting often exist in a feminine sphere?"**

Shayna Connelly: "There's a link between hauntings and the cultural gaslighting of women. It's well documented that women's physical pain is dismissed as being 'all in her head,' and with the stereotype of the 'hysterical woman,' it's easy to see how emotional pain would also be dismissed as imaginary. Psychosomatic pain, which by definition originates in the brain rather than the body, is still pain. It hurts. Feeling pain of any sort and being told it doesn't exist create a haunting presence in a woman's life. It's important to add that this gaslighting is even more extreme for women of color and transgender folks, who face even greater hurdles and abuses trying to have their experience acknowledged and pain treated."

Kelly: **"I hadn't thought about the physical pain part of this."**

Shayna Connelly: "Hauntings are linked to trauma, and anyone who is not part of the dominant culture is traumatized in a myriad of ways daily. It's important to differentiate between ghosts and hauntings. Ghosts are linked to belief that the essential part of ourselves continues to exist after our body dies, which is a belief not everyone shares and can be debated. There is a metaphoric use of the word 'ghost' to talk about people on the margins of society who are invisible, and that includes the mentally ill. Hauntings affect us all regardless of our belief in souls and afterlives. Hauntings are not necessarily linked to the spirit of a dead person straddling the line between life and afterlife. A liminal space—a state of becoming between two states that we typically think of as either/or (such as alive or dead, child or adult) is an aspect of hauntings. Hauntings are a natural outgrowth of trauma and are universal human experiences. We are personally traumatized by heartbreak, injury, violence, or death. There are collective traumas that affect large groups such as war, terrorism, displacement, imprisonment, or poverty."

Meg: **"What about the art of experimental film allows you to create in a unique way? Do you think it allows exploration of feminism and horror more than traditional cinema?"**

Shayna Connelly: "Experimental cinema is as varied as narrative cinema in terms of forms and genres. A lot of experimental work deals with associations between images in lieu of narrative cause and effect. It functions more like poetry than prose. Narrative's cause and effect are tied to the natural, rational world, where there is a scientific explanation for what we witness. Our senses are notoriously limited, and yet we think of them as being all-inclusive, omniscient windows to the world. So much information is obscured from us and then additionally filtered through our beliefs, conditioning, and emotional states. This makes our empirical understanding of the world primarily associative rather than causal, yet we frame it as causal. I suppose I reject this causality as too simplistic, though I also have trouble with plot-heavy work. It excludes so much nuance, and experimental work is all about nuance and truths that cannot be expressed in words. Association lends itself well to a contemporary female or nonbinary experience, since women and nonbinary folks are often unrepresented or falsely represented in the narrative mainstream. Going back to my early obsession with ghost photography, the image of 'The Brown Lady' is an image that stuck with me, but it is an appropriated image. Found footage filmmaking, a tradition that appropriates and recontextualizes footage, lends itself to exploring beauty and horror simultaneously."

Meg: **"I like this concept of an obsession with one image and how it translates into film."**

Shayna Connelly: "There's a myth that experimental forms are impenetrable, but they're like poetry and function on impression, association, or dream-like logic. There are enough experimental forms that when someone is exposed to the breadth of work, there is something for everyone. Narrative is so ingrained in us that it takes a while to be able to "read" nonnarrative form, allowing yourself to sink into and just exist with what the film has to offer rather than looking for character arcs and causality. Because the supernatural is beyond causality, it is already more experimental/feeling/poetry than concrete, logical narrative reality. What interests me is that I often find people who don't consider supernatural stories a part of horror. Horror is about transgressing boundaries. Body horror's revulsion comes from seeing things that should be contained inside the boundary suddenly visible outside the body. Ghosts, vampires, and zombies all transgress the boundary between life and death, though vampires and zombies are concrete physical presences, whereas ghosts lack physicality and their presence is often in question."

Kelly: **"Mental illness and the supernatural have long been intertwined in the Female Gothic tradition. Why do you think this is? And why has it become an integral piece of your storytelling?"**

Shayna Connelly: "Because ghosts-as-monsters are beyond the realm of the natural world, there is always a question of whether they are truly external forces or exist as part of a psychosis. Mental illness and the supernatural are immediately intertwined with death. Grief renders a person insane. It's temporary insanity, but no less pathological. Grief and other traumas interfere with a person's ability to reason, they alter our perceptions, all of which distances us from the world. We see, hear, and experience things that are not accessible to the people around us. That estrangement is a netherworld we exist in for a time, and for the sufferers this subjective state is the only reality they know. In this state our perceptions can make us feel pursued by a violent, invisible entity. We may feel possessed, we may feel the presence of someone we loved who is gone. These lived experiences must be acknowledged as real, but particularly in women these reactions are dismissed as insanity. Healers are needed to guide a suffering person through the grief or trauma and out the other side. Whether that help is best from a spiritual leader of a community, a psychologist, or a medical doctor, I don't know. Even when we start to heal and reintegrate with the world, the self is permanently changed. There's a link between the haunted female protagonist and mental illness, which I first saw in *Let's Scare Jessica to Death* (1971) when I was about eight. The protagonist had been previously committed and starts seeing ghosts in her new environment. The question of whether the ghosts are real or a product of female insanity is central to the plot. Other films like *Images* (1972) also work with this idea. Culturally we're obsessed with madness and dismissing madness, but being female is one of the identities in the world that is chronically traumatizing."

Meg: **"Please tell us about your current projects and where we can see your work next!"**

Shayna Connelly: "My ghosts have shifted in two directions. I have two short screenplays exploring the idea of the 'living ghost,' someone who is disappearing, but still alive. Someone who is aging is in a liminal state similar to adolescence, but the transition out of adulthood seems to be death. *Love as Practice for Dying* is about a man preparing for his death while haunted by the estrangement from his daughter. *The Tensile Strength of Air* is in preproduction and deals with a woman coming to terms with her changing identity in middle age. A woman's status in the world changes dramatically as she ages, and it is difficult to understand, despite having a lot of clear benefits. I see both films as horrifying because the abjection in them stems from the disappearance of the characters' previous identities and needing to live in a world where they are only partially seen. The other direction I'm exploring is a change in form and substance. I've jettisoned the nonscary ghost for a malevolent, haunted world in a feature screenplay that's currently bonkers-experimental in form. Describing something that's only partially formed like this script is like trying to grasp air. The story is in three parts with the same characters in altered states of being and each part bending the rules of causality differently. It deals with the boundaries of individual versus shared grief and how sharp the edge separating reality and madness is. It has nightmare imagery scattered throughout—things that at first glance seem familiar or benign, but after looking longer something is not as it should be."

Kelly: **"We can't wait to see more of your work!"**

Speaking to a filmmaker about her work on the very topic of female hysteria enlightened us further about gothic storytelling. Womanhood is clearly a complex topic rife with possibility in the horror genre. We're excited to discover more about how science and film interact.

SECTION NINE
WOMEN IN A MAN'S WORLD

CHAPTER TWENTY-FIVE
GHOSTBUSTERS

Year of Release: 2016	
Director: Paul Feig	
Writer: Katie Dippold, Paul Feig	
Starring: Melissa McCarthy, Kristen Wiig	
Budget: $144 million	
Box Office: $229.1 million	

Horror, like many genres, is prone to sequels and remakes. The original *Ghostbusters* (1984) spawned the popular animated series *The Real Ghostbusters* (1986–1991) as well as the sequel *Ghostbusters II* (1989). More than that, *Ghostbusters* became a cultural phenomenon in the form of toys, cartoons, and films, a zeitgeist that rarely occurs. Talk of a third film had been a popular topic online among "ghost heads" and casual fans alike. That is, until Harold Ramis, cowriter and Egon in the films, passed away in 2014. Then, it became clear that the hope of the canonical four appearing on-screen again would never materialize.

In the original *Ghostbusters*, a charming foursome of ghost hunters (Harold Ramis, Dan Aykroyd, Bill Murray, and Ernie Hudson) rid New York City of evil, all while dropping memorable one-liners. While the 1984 *Ghostbusters* still holds up, continuing to be delightfully fun,

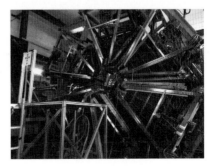

MIT particle physicist James Maxwell helped redesign proton packs, the movie's famous ghost-catching tool, to bring it more in line with modern particle accelerators such as the Large Hadron Collider.[1]

a modern interpretation can see the cracks in its treatment of women. There are barely any female characters, and Peter's (Murray) advances toward Dana (Sigourney Weaver) are antiquated in their aggressiveness. Dana, though rather headstrong, is fated to be invaded by the demon, Zuul, and becomes the damsel in distress. Zuul uses his powers to turn Dana into a vixen, a direct fulfillment of how Peter wanted her to be.

Given that there were improvements that could be made for the modern age, director Paul Feig came aboard to reboot *Ghostbusters*, much how popular franchises (*Halloween* in 2007, *Friday the 13th* in 2009) had in recent years. Yet, no one could have predicted the backlash the reboot, starring four women, would suffer:

> In March 2016, fan outrage reached an apex in the days following the online debut of the *Ghostbusters 2016* trailer, which swiftly became the most disliked film trailer in YouTube's short but impactful history. Online news media picked up on the story and orchestrated a cultural firestorm, primarily hinged on a minority cluster of misogynist comments.[2]

This also manifested in a planned effort by fans of the original film to lower the rating of the reboot on its IMDB page, before they had even seen it. Of these one star reviewers, there were eight times as many men as women. It is difficult not to see that gender is at the root of the vitriolic hate directed at the movie, Feig, and even its four female stars (Leslie Jones, Kate McKinnon, Melissa McCarthy, and Kristen Wiig). Jones, an African American actress known for her work on *Saturday Night Live* (2014–2019), received the brunt of the verbal attacks. The social media abuse was so severe, including strangers sending her pornography and calling her racial slurs, that Jones eventually deleted her Twitter account. So, what angered fans so severely that it led them to this extreme reaction?

> Is there anything to the backlash beyond anger that someone took down the "No Girls Allowed" sign outside the Ghostbusters clubhouse? Can a remake, reboot, or sequel actually harm the original? . . . So what changes, and what's at stake? The movie

is immutable, but time marches on, and we are borne along with it. . . . For (some) the new Ghostbusters is a long overdue vindication of the idea that you don't need to be a man to strap on a proton pack.[3]

In "Why the Ghostbusters Backlash is a Sexist Control Issue" for *Indiewire*, Sam Adams continues, telling the story of James Rolfe, an angered fan of the original:

For Rolfe, the new movie's very existence is a blot on the original, a permanent asterisk next to its name. "I hear all the time, the female Ghostbusters," he says, "does that mean we have to call the original the male Ghostbusters?" Intentionally or not, Rolfe's complaint cuts to the heart of the matter (The Ghostbro's lack of self-awareness is a gift that never stops giving). We've long had the habit of using the universal to refer to men while shunting women into their own subcategory.[4]

Ectoplasm (from the Greek *ektos*, meaning "outside," and *plasma*, meaning "something formed or molded") is a term used in spiritualism to denote a substance or spiritual energy "exteriorized" by physical mediums. It was coined in 1894 by psychical researcher Charles Richet.

This concept of women's horror being forced into a niche space is one founded on the misnomer that we do not enjoy, or seek out, the genre. At the 2017 remake of *It*, gender lines were split evenly, as 51 percent of ticket buyers were men, and 49 percent women.[5]

The women of *Ghostbusters* are fully realized heroes, females with brains and brawn. In her article for the *Washington Post*, Petula Dvorak argues that it is the film industry's dearth of female representation that is partly to blame for the negative publicity:

Remember, women accounted for only 30 percent of speaking roles in the top one hundred films of 2013, according to *It's a Man's (Celluloid) World* survey by Martha M. Lauren, executive director of the Center for the Study of Women in Television and Film at San Diego State University. That study also showed women were protagonists in only 15 percent of the movies.[6]

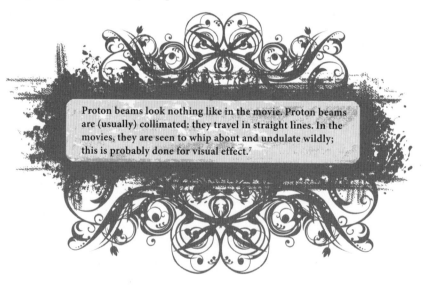

Proton beams look nothing like in the movie. Proton beams are (usually) collimated: they travel in straight lines. In the movies, they are seen to whip about and undulate wildly; this is probably done for visual effect.[7]

These egregious percentages are what inspired actress Geena Davis to found the Geena Davis Institute of Gender in Media in 2004. The mission of her institute is to balance the disparity of the genders in all aspects of media, particularly that which is consumed by young children. To solve this overwhelming problem, the institute has developed a rubric, GD IQ (Geena Davis Inclusion Quotient) to grade films and TV series. Using a computer algorithm, the quotient measures screen time and speaking lines for male and female characters. Months after the *Ghostbusters* controversy, the GDI published the findings of their first study using this quotient. They used the two hundred highest grossing nonanimated films of 2014 and 2015. The results were significant and point to the notion that no matter the genre, Hollywood is still male-dominated. First, "male characters received two times the screen time as female characters in 2015. And, when a film has a male lead, the gender gap is even wider, with male characters appearing on screen nearly three times more often than female characters."[8] These numbers were the same when analyzing speaking lines, and, most unsettling, in "films with female leads, male characters spoke about the same as female characters."[9]

The good news is that horror is the most robust genre for female-driven stories. In a recent study, women appeared in 53 percent of the screen time and 47 percent of the speaking time. Films like the 2016 *Ghostbusters* aid in the disparity of genders, a reality even in 2020. And thanks to foundations like the Geena Davis Institute, women, and women of color, are finding their voice both in front of and behind the screen. We talked to filmmaker Samantha Kolesnik about life as a female director in the horror genre, and whether she feels as if she were in a "man's world."

Meg: **"First, can you tell us about your first realization that you wanted to create horror content?"**

Samantha Kolesnik: "I remember the first time, as a youth, that I read *The Cask of Amontillado* [1846] by Edgar Allan Poe. There's so much about the dark side of the human condition that's conveyed in that story. Poe showed me that humans can be scarier than werewolves and witches. What's more terrifying than considering that man is actually the apex monster? Basically, human nature and human history aren't all rainbows and dandelions. And that's explored to great effect through horror."

Kelly: **"Have you ever felt that you were infiltrating a 'man's' genre?"**

Samantha Kolesnik: "No, I have never felt as though I were infiltrating because I do not view horror as a 'man's' genre. There are cultural structures in place that try to, and have historically tried to, make it a man's genre, but horror is just as much 'my genre' as it is anyone else's. At the end of the day, I retain the power to tell my own stories how I see fit."

Meg: **"Have there been people in the industry who've attempted to make you feel this way?"**

Samantha Kolesnik: "Sure. There are systemic and individualized attempts to disempower women. It can range from a man telling you that you won an award because you're a woman, rather than due to the quality of your work—all the way to an entire event's culture not being welcoming to women filmmakers. I've experienced both and more. I'm not interested, at this point in my life, in laundry-listing the ways others have sought to diminish me. I'm interested in building, empowering, and moving forward."

Kelly: **"In *Mama's Boy* (2018), you explore the relationship between mother and child. Why was this an important theme for you?"**

Samantha Kolesnik: "*Mama's Boy* is about trauma. The horror of *Mama's Boy* is that, despite the violent acts our main character, Joshua, commits toward others, the most disturbing are the ones he inflicts upon himself in an endless cycle of recreating the abusive relationship he had with his mother. Since he tries to leave his mother behind by vanquishing those who have debased him, he returns to her in the end. She, and the pain she has inflicted, never leave. It's his only constant in life, his true companion."

Meg: **"Tell us about Above the Line Artistry. What is the aim of your production company? What sort of stories do you want to tell?"**

Samantha Kolesnik: "Above the Line Artistry was founded by Vanessa Ionta Wright, a collaborator and dear friend of mine. She brought me on as a coowner later, a couple of years ago. The company was created initially to produce the film *Rainy Season* (2017), which is based on the short story by Stephen King. Vanessa and I really hit it off during that production and decided to keep working together, and to grow the company. Our focus has been on horror, but that may change moving forward. We're not defined by a single genre."

Whichever *Ghostbusters* you prefer, this chapter and the GD IQ demonstrate that inclusion is better for everyone.

CHAPTER TWENTY-SIX
THE X-FILES

Years of Production: 1993–2018	
Created by: Chris Carter	
Starring: Gillian Anderson, David Duchovny	
Network: Fox	

The pilot episode of *The X-Files* begins with a nameless woman being chased in the Oregon woods. Cut to daylight, and there is a group of stern-faced men surrounding her lifeless body, speculating over whether she'd been sexually assaulted. This imagery of female victim and male law enforcement is familiar. The nineties saw a boom of this juxtaposition in films like *The Fugitive* (1993), *Kiss the Girls* (1997), and the rise of the *Law and Order* (1990–2010) franchise. One such significant film garnered multiple awards and audience adoration: *The Silence of the Lambs* (1991). Instead of the presumed male detective, Clarice Starling (Jodie Foster), a relative rookie, is the film's protagonist. From the beginning, we see Clarice's navigation through the male world of the FBI, to the male world of the prison where Hannibal Lecter (Anthony Hopkins) resides. She literally has to avoid the semen flung by inmate Miggs (Stuart Rudin) in order to do her job. Later, when Clarice is in a small police station with her boss (Scott Glenn), he and the other males in the room worry she cannot handle the task of surveying a body for evidence in the tight, male-populated space. The sexism is palpable, as Clarice is constantly put in the same, helpless category as the victims of Buffalo Bill (Ted Levine). In *Manhunting: The Female Detective in the Serial Killer Film*, Philippa Gates examines this phenomenon:

> Following the success of *The Silence of the Lambs*, Hollywood film saw an increasing presence of female detectives on screen. This

shift away from white hypermasculinity would suggest a more liberal and feminist approach to the definition of law enforcement heroism. However, while the detective genre has brought women to the center of the narrative with a seemingly greater degree of agency as the protagonists who drive the narrative action forward, this agency is tempered and contained. The male detective is empowered in the contemporary detective film through his identification with the serial killer—the man who has the desire and ability to inflict violence on women—while the female body remains a site of objectification and powerlessness."[1]

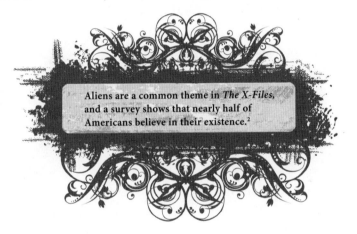

Aliens are a common theme in *The X-Files*, and a survey shows that nearly half of Americans believe in their existence.[2]

Dr. Dana Scully (Gillian Anderson) of *The X-Files* is an obvious descendant of Clarice. Born to the screen just two years later, Scully is styled in an almost exact likeness. Wearing sensible shoes and a no-nonsense suit, we meet the medical doctor turned FBI recruit immediately after the female victim's scene. And just as director Jonathan Demme showed Clarice in male-dominated halls, Scully is shown surrounded by males as she makes her way to an office. An office stuffed with men, both in front of and behind her. Next, we follow Scully downstairs to the basement, where she meets yet another man, this one known to be rebellious and "spooky." Tasked with essentially babysitting Fox Mulder (David Duchovny), Scully quickly becomes the voice of reason. She brings the logic and science, while Mulder takes on the more traditionally female trait of emotionality in his staunch beliefs of the supernatural. Beliefs that sprouted from a past trauma. While this role reversal is refreshing, particularly for the era, it leaves us to ask that if Scully were the rebel, the one with a spooky reputation, would she even be allowed to work at the FBI? Or is it the fact that she shows masculine traits that aids in her success both in the sphere of the FBI and as a character?

Thankfully, *The X-Files* became a hit, and Scully was given decades to grow into one of the most complex, authentic, and important females in television history. So important, in fact, that the aforementioned Geena Davis Institute of Gender in Media coined the phrase "Scully Effect" in their 2018 study of media's influence on women's career choices. In the study, several groups of women were tested: those aged twenty-five and older who were old enough to have watched *The X-Files*; women actively employed in science, engineering, math, or technology (STEM); and women who identified as *The X-Files* fans. The study led to many fascinating findings including that half of those women who were familiar with Scully say she increased their interest in STEM. And, that medium/heavy female viewers of *The X-Files* were 27 percent more likely to have studied STEM than non/light viewers. Also, 63 percent of women familiar with Scully said that she increased their own confidence that they could excel in a male-dominated profession.[3] As the Geena Davis Institute's catch phrase "if she can see it, she can be it" suggests, these significant findings show that media are an integral part of our lives. When creatives and studios invest their time and resources into showcasing women like Scully, society benefits as a whole.

The Scully effect inspired girls and women to pursue STEM careers.

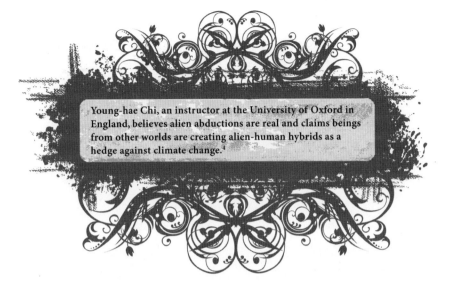

Young-hae Chi, an instructor at the University of Oxford in England, believes alien abductions are real and claims beings from other worlds are creating alien-human hybrids as a hedge against climate change.[4]

One industry that still remains dominated by men in 2020 is television and film production. While *The X-Files* gave us Scully, it was also overwhelmingly written, directed, and produced by men. The aim of women's increased role in front of the camera is important, yet their inclusion behind the camera is equally vital. In 2018, horror producer Jason Blum notoriously declared that "there are not a lot of female directors period, and even less who are inclined to do horror."[5] This prompted immediate outrage, to which Blum publicly apologized. His claim is obviously refuted by the contents of this book, as we've had the occasion to interview female horror directors. Yet, the misconception remains.

We imagine that Dana Scully, forever a role model, would find success in whatever field she chose. Because of her timely appearance on television, at the rise of the women's workforce, and because of the "Scully Effect," we have more women in STEM than ever before. This is a powerful lesson in the gravity of media, and how it is our responsibility to foster multitudes of relatable, ambitious, and complicated female characters.

CHAPTER TWENTY-SEVEN
GINGER SNAPS

Year of Release: 2000	
Director: John Fawcett	
Writer: Karen Walton, John Fawcett	
Starring: Emily Perkins, Katharine Isabelle	
Budget: $4.5 million	
Box Office: $572,781	

When the term "movie monster" is used, the somber faces of Universal's iconic monsters naturally come to mind. From the 1920s to 1950s, Universal Studios created the framework for modern horror. Starting with *The Hunchback of Notre Dame* (1923) and *The Phantom of the Opera* (1925), both starring Lon Chaney, Universal began a male-dominated run of films where the men were the monsters and the women were the prey. This trend continued with hits *Dracula* (1931), *Frankenstein* (1931), *The Mummy* (1932), *The Invisible Man* (1933), and *The Wolf Man* (1941). The only female creature from these spate of movies to transcend the era and become an icon is the Bride of Frankenstein. Considered by many to be better than the original, *Bride of Frankenstein* (1935) was the first sequel to *Frankenstein*. It centers on a subplot in the 1818 novel by Mary Shelley in which the creature longs for a mate to be created by Dr. Frankenstein. Portrayed by Elsa Lanchester, the Bride has little screen time in the film, but her aesthetic has endured, including her tall, conical hair-do. Unfortunately, at the end of *Bride of Frankenstein*, the creature, feeling rejected by his intended bride, sets the castle ablaze. This scientific and masculine rejection of the female seems like an appropriate end, similar to the themes Mary Shelley explored in her novel, and so she would, perhaps, approve.

Special effects makeup for *Ginger Snaps* took five hours to apply and two hours to remove each day for filming.[1]

Praised as one of the top emerging directors in the horror genre, Gigi Saul Guerrero has explored many aspects of female horror. We were lucky enough to speak to the recent director of the timely film *Culture Shock* (2019).

Meg: **"You have many roles in filmmaking (actress, director, writer). Was this your dream as you studied at Capilano University in their motion picture degree program? Or does taking on multiple roles happen naturally as you immerse yourself in making movies?"**

Gigi Saul Guerrero: "Still to this day I don't know if I prefer one or the other. I started out in acting, mainly theater and musical theater before film school. Once I got an agent in my teen years and started booking commercials and other roles, I really observed how other directors handle the set. It was not until film school that I really learned how different yet how much actors and directors depend on each other. I mean . . . if I could be the less crazy version of Mel Gibson, I would be the happiest and do both! But in all seriousness, I have enjoyed so much over the years taking my turn in each role, I find I learn so much for all of them and I actually apply it to the next project. Working as an actress I have really watched for what to do and not to do as a director. Same when it comes to writing, coming from an actor's point of view.

I have really been careful in writing my scripts to be clear as day on what my intentions are. I am loving every second that all these different roles I like to do as a storyteller play a very important part in the creation of visual stories."

Kelly: **"Horror film has been criticized in the past for its treatment of women, particularly women of color. As a Mexican-Canadian filmmaker, can you describe why it's important for you to create films that feature and highlight your culture?"**

Gigi Saul Guerrero: "I think being female in general in this industry, it is important to be true to yourself and recognize who you are as a storyteller. What makes you different? What makes you stand out from the overpowering crowd? In my case, I'm Mexican. Living in Canada never stopped me from sharing my culture out loud. It's what makes me *me*! So, in terms of filmmaking, I want to create what I know and what I am comfortable sharing. I think it's so important for any filmmaker starting out to find their personal voice and stick to it until it's time to move on to different obstacles in film. I never see it as a disadvantage being female or of color in the horror community or even in the film industry. I find that a waste of time if I think that way. Instead, I look at it as an advantage to work harder, to push myself to create what really matters to me. At the end of the day, I want to be recognized as a *great* filmmaker, no other label attached."

Meg: **"Your feature film debut *Culture Shock* that you made with Blumhouse Productions and Hulu has really made a stamp for its horrific realism of the border crisis we are facing. What do you have to say about the timeliness of the film and your outlook on what's going on at the border? Did you expect to make something like this?"**

Gigi Saul Guerrero: "I knew when I first got my hands on this script this was something I can tackle and *want* to tell. It felt very personal to me. It's amazing how the studios of Hulu and Blumhouse trusted me to tell this story, and an even bigger win that they wanted to be part of such an important story. The brutality and the inhumane subject matter is there, and unfortunately, we see that every day, for a while now, and it's quite heartbreaking. I didn't want to make something so heartbreaking that we'd have trouble watching—because we already have enough trouble watching it every day on the news. Instead, I wanted to create something entertaining, with drama and humor so we could escape the horror realities we face. Unfortunately, it is really timely. But in another way, I feel it's time to really recognize what is happening at the border and in our world in general. It's hard to explain because it's such an angry situation and really hard to put words together about it. I hope that this film at least begins a conversation. It's very hard to help what's happening, but if we can at least talk about it, and at least recognize how awful the situation is, we're taking a big step forward."

Kelly: "**Could you tell us about your upcoming projects and where we can see your work next?**"

Gigi Saul Guerrero: "Really looking forward to all the amazing opportunities my first feature has brought me. I am directing for TV and pitching in the room often for a theatrical release! It's a blessing how fast-paced this industry is. By staying focused, humble, and dedicated, you would be surprised how your dreams start coming true. My company, Luchagore Productions, and I have big plans! If you follow myself or Luchagore on social media, you will find all the amazing gory fun we are creating!"

Women killers run rampant in horror films, particularly those of the psychologically damaged variety like Annie Wilkes (Kathy Bates) or Pamela Voorhees (Betsy Palmer). Women creatures are far less prolific, often relegated to sexy vampires or Victorian ghosts.

In the Canadian film *Ginger Snaps* (2000), women invade the masculine world of werewolves. Traditionally, lycanthropes in film have been men pursuing their primal urges at the call of the full moon. Often, they are males who could be described as weak, like Scott (Michael J. Fox) in *Teen Wolf* (1985), who relies on his transformation to win a basket-ball game. In *An American Werewolf in London* (1981), David (David Naughton) is bitten by a werewolf on the mysterious moors of England. As Elizabeth A. Lawrence writes in *Werewolves in Psyche and Cinema: Man-Beast Transformation and Paradox*, David's transformation is in line with many cinematic masculine werewolves before him:

> Conforming to the familiar pattern, David makes love to his nurse, who has fallen for him. He is shown urinating. A frightened cat spits at him. On the next full moon, he tears off his clothes, revealing a hairy body; he develops fangs and paws with long nails. He feel invigorated and his body seems very strong.[2]

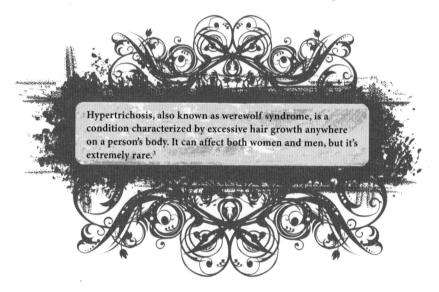

Hypertrichosis, also known as werewolf syndrome, is a condition characterized by excessive hair growth anywhere on a person's body. It can affect both women and men, but it's extremely rare.[3]

After Ginger (Katharine Isabelle) in *Ginger Snaps* is attacked by a werewolf, she exhibits strange behavior. This includes her almost immediate inclusion into a group of male students, an appropriate symbol of her joining a boys' club of sorts. This escalates, as she leaves her sister, Brigitte (Emily Perkins), behind to spend more time with males, going against her usual preference. One night, Ginger, newly in tune with her sexuality, spends the evening in the car with Jason (Jesse Moss). As they kiss, and Ginger's aggression heightens, Jason asks her, "Who's the guy?"

This is to indicate that Ginger has subverted her feminine role, rebelling against how society believes she should act in a sexual encounter. Instead of demurring, Ginger becomes more aggressive. Afterward, Ginger attacks and presumably ingests a neighbor's dog. When she returns home, Brigitte sees the blood and assumes Jason has hurt Ginger, when the opposite is true.

It is vital to note that Ginger's transformation happens on the exact day of her first menstrual period, furthering the symbolism of her womanhood:

> The Fitzgerald sisters are told by their school nurse that Ginger's changing biology, which includes heavy bleeding and sudden hirsutism, are all perfectly normal and come "with the territory" of female sexual maturity, cheerfully predicting similar monthly visitations for the next thirty years. While Ginger breaks the mold in that her lycanthropy is not cyclical but rather progressive and irreversible, the binding Brigitte uses to conceal Ginger's tail calls to mind the elaborate paraphernalia of archaic feminine hygiene products, while the girls' furtive anxiety at being found out mimics common female fear of "spotting."[4]

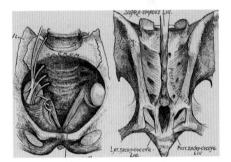

Humans possess vestigial tail muscles. There have been more than twenty recorded cases of humans born with tails since the 1800s.[5]

For Ginger, her body changes into something inherently male rather than in the "normal" way a teen girl should grow. This subversion is another form of rebellion against a patriarchal society: "The female werewolf embodies a kind of gendered body crossing: a female body expressing characteristics labeled both masculine and male by the dominant culture (power, strength, rage, aggression, violence, and body hair)."[6] Unable to control the dramatic transformation to her body, Ginger succumbs, and like David and Scott, she embraces the primal advantages of being a wolf. She no longer displays the female tendency of adopting guilt and shame, fully giving in to the sexual undertones of her kills. This sexual primacy of a female creature is also on display in the Karyn Kusama film *Jennifer's Body* (2009) in which succubus, Jennifer (Megan Fox), uses her sexuality to feed on unsuspecting teenagers. Alien-human hybrid Sil (Natasha Henstridge) of *Species* (1995) holds a similar sexual pull over the men in her orbit.

Like David in *An American Werewolf in London*, Ginger pays for her complicity in succumbing to her primal side. She is killed, destined to forever be encased in the body of a wolf. This fate echoes in female creatures like Jennifer and Sil, who both are destroyed at the end of their respective films. Yet, something they are able to achieve is a "procreation." Like archetypal mothers, Jennifer and Sil live on in their offspring. For Jennifer, she has passed on her succubus trait to friend Needy (Amanda Seyfried), and for Sil, her genes are carried on through a rat who chews on her severed tentacles. This is true for Ginger, too, as her sister, Brigitte, carries on the female lycanthropy in *Ginger Snaps 2: Unleashed* (2004). Let these examples be a lesson that female creatures may just have a better staying power than their male counterparts.

SECTION TEN
KICK-ASS WOMEN

CHAPTER TWENTY-EIGHT
BUFFY THE VAMPIRE SLAYER

Years of Production: 1997–2003	
Created by: Joss Whedon	
Starring: Sarah Michelle Gellar, Alyson Hannigan	
Networks: The WB, UPN	

Males who watched sexually violent shows with submissive female characters reported more negative attitudes about women than the control group. This effect did not occur for men who watched shows with powerful women. . . . Women who watched weak characters in sexually violent situations became twice as anxious as women who watched *SVU* [1999–present] or *Buffy* [1997–2003], who in turn actually reported less anxiety than the control group. The inverse occurred for men, who felt least anxious after watching *The Tudors* [2007–2010] or *Masters of Horror* [2005–2007].[1]

This theory is a driving force behind the Geena Davis Institute on Gender in Media and their slogan, "If she can see it, she can be it." How women, and all people, are portrayed in media can affect children especially, who are consuming up to seven hours of media per day. Seeing women, people of color, and those with disabilities portrayed in a variety of roles can have a positive impact on consumers.

One television show that gave a woman the chance to lead was *Buffy the Vampire Slayer.* "For TV scholars, *Buffy* was the birth of what we now call 'quality television,' in terms of shows that hit certain characteristics and that we can discuss as a quality text," Buffy studies scholar Lorna Jowett said. "We learned how to talk about television as an art form from this show. We're still talking about whether it's feminist many years later."[2]

Buffy creator Joss Whedon based the character on the archetypal blonde girl who seems to get killed first in every horror movie. In his DVD commentary for the first episode, he says, "The idea of Buffy was to subvert that idea, that image, and create someone who was a hero where she had always been a victim." Buffy was first portrayed in a movie version in 1992 by Kristy Swanson and then moved to the small screen a few years later with Sarah Michelle Gellar as the title character.

There is a trope in film and television that depends on a belief that:

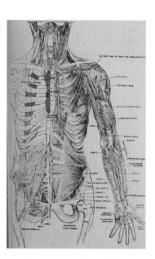

Buffy is trained to fight throughout the series. The world's strongest woman, Becca Swanson, is currently the only woman to have squatted 854 pounds and bench-pressed 600 pounds.[3]

. . . the male action hero is more important than anyone else in his story, and that any pain and misery felt by other characters is subordinate to his. In stories that are built on these tropes, female characters do not have worth in and of themselves; they are valuable only insofar as they can add to the male hero's pain. Moreover, the male character's pain is unique. . . . . Action stories are disproportionately built around (white, usually straight, usually wealthy) male characters, so their pain is disproportionately privileged over that of other characters. Female characters are disproportionately tortured and killed as part of stories about male characters, instead of being granted their own stories. It's not the individual stories that are troubling so much as it is the overwhelming accumulation of them. They send a message, and that message is: Here are the characters who deserve your empathy. Here are the people in the world whose pain really and truly matters. They are (white, straight, wealthy) men.[4]

Buffy subverts that idea by having the audience relate to *her*. She's not big; her small stature and slight frame make her seem like an easy target.

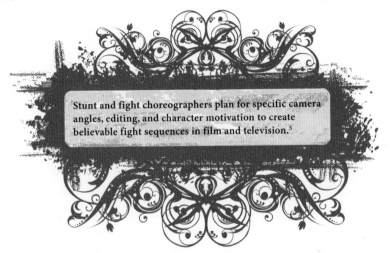

Her slayer powers, of course, give her that extra advantage to fight off any baddies that may try to do her harm. She's also not perfect. She makes mistakes, makes bad decisions, and cracks jokes when she should be taking something seriously. This made her more relatable to the audience and, in turn, more believable.

Stunt and fight choreographers plan for specific camera angles, editing, and character motivation to create believable fight sequences in film and television.[5]

Buffy also doesn't necessarily want to be the hero. She would rather live a normal teenage life and not have to worry about the undead. She accepts her calling, though, and is trained by her watcher Giles (Anthony Stewart Head) throughout the first several seasons. She learns to be focused, quick, and studies combat skills. Buffy never fails to throw in a quick one-liner, too, while fighting her monster of the week.

One of the most touching and realistic episodes of *Buffy* is "The Body." This season-five episode was written and directed by series creator Joss Whedon and contains no musical score. We watch as Buffy discovers her mother's dead body and proceeds to try to resuscitate her. In what feels like real time, we wait for the ambulance to arrive with Buffy and feel helpless with her. These moments of showing a real, vulnerable young woman go through unimaginable grief allow us to connect with the character and keep her grounded. Buffy may be physically strong, but it's her inner strength that truly makes her a hero.

The audience traveled with Buffy through some very real, albeit sometimes supernatural, emotions and situations. We saw her handle breakups, rejection, death, depression, feeling inadequate, and having to take on more responsibility than she thought she was capable of. Was she always perfect? No. Did she make mistakes? Absolutely. Having Buffy be a teenager and then a young adult going through real situations helped the audience engage and relate even more to her during the otherworldly trials she faced.

Other female characters on *Buffy* include Willow (Alyson Hannigan), Cordelia (Charisma Carpenter), and Anya (Emma Caulfield). Each has their own unique character traits and, as we discover in later seasons, otherworldly powers. Willow is the book-smart, loyal friend who may not be physically strong but will stand up with and fight for her friends. In later seasons, she develops into a powerful witch and even becomes the antagonist of the series after her girlfriend Tara (Amber Benson) is senselessly murdered. This decision to kill off a beloved character had fans reeling, and many blamed showrunner Marti Noxon for the choice. Even though she may have received a lot of fan anger, the seasons she oversaw contain, arguably, some of the best episodes of the series. (Anyone else ship Buffy and Spike?)

One of Buffy's stunt women, Michele Waitman, earned a scholarship to East Tennessee State University for gymnastics and cheerleading.[6]

Noxon became showrunner of *Buffy the Vampire Slayer* during season six, and markedly, it's the season that shows Buffy going down some darker paths and character development. Noxon said:

There's so many theories about why we like to watch these stories about women being complex and making mistakes and, I can only say what the answer is for me, which is that there's a real catharsis in seeing women be the people with agency in their stories, women who are committed to the full range of emotions. . . . We have all the same feelings as any other human being. We can be completely shitty—just like a man. And we don't necessarily have to have a really sympathetic backstory.[7]

The relationship between Willow and Tara was groundbreaking in its portrayal of a lesbian couple on television. To learn more about queer representation in the media, we spoke to Melody Lynn Hoffman, who teaches at Anoka-Ramsey Community College in Minnesota.

Kelly: "**Could you tell us about your educational background and what made you first interested in queer studies?**"

Melody Lynn Hoffman: "I have a PhD in Communication Studies with an emphasis in feminist and sexuality studies from the University of Minnesota. I received my bachelor's and master's degrees in Journalism and Mass Communication at the University of Wisconsin-Milwaukee. I was hooked on representational analyses the moment I read the first line of my undergrad gender and media textbook. That's a bit hyperbolic, but you get my point. I simply love analyzing identity representation in media texts. My representational research is intersectional with a focus on gender and race. What I love and admire about queer studies is the subversive readings done within media texts. I do not love that these subversive readings are historically a necessity, but I am always interested in oppositional readings of dominant messages. An example of this is the US gay community's reading of *The Wizard of Oz* (1939). Judy Garland is known as a gay icon because gay men related to her story line and thus called themselves "friends of Dorothy." A dominant or heterosexual reading of the film would not see this oppositional reading. *The Celluloid Closet* (1995) is a major research inspiration for me. The documentary and book explore how filmmakers snuck in queer characters during the Hays Code Era. That research taught me that filmmakers have always been queering media texts, even when homosexuality was forbidden in US films."

Meg: "**Are you a fan of any horror media? How do you think they stack up against other genres in the depiction of queer characters?**"

Melody Lynn Hoffman: "Horror media have always intrigued me. As a kid, I read R.L. Stine's *Goosebumps* book series (1992–1997) and was a loyal watcher of *Are you Afraid of the Dark?* (1990–1996). *Scream* (1996) was a huge movie when I was a young adult, and it taught me all about horror film tropes. Given my lived experiences and how I approach media, I do not watch every horror film that hits the theaters. I tend to favor complex horror films like *The Cabin in the Woods* (2011) and *Get Out (2017).* A few years ago, I decided I wasn't going to waste any more of my time watching predictable and uncreative media made by old white straight men. White men control and produce almost all of the films we see in US movie theaters, so I put my money and time toward all the other people making films. I miss out on a lot of the mainstream horror films, but I feel ok about that. Directors such as Joss Whedon (the exception to the white man rule) and Jordan Peele make creative films that push against representational stereotypes. I am all in for that."
Kelly: "**So are we!**"

Melody Lynn Hoffman: "Queer communities in the US do have a particular affinity for horror films. One reason is that horror films often contain characters ostracized from mainstream society, and queer audiences can certainly identify with that. Second, horror films have long contained queer characters, and so it is a film genre that queer people can find themselves in—even if the representations are problematic. Before I dig into specific queer representations in horror film, I want to make something very clear. Queer representations are not either here or not here—they can be simultaneously both. Queer characters often exist in between the lines. A character does not need to be introduced as queer to be read as queer. Queer audiences have always searched for characters they could relate to, even if the filmmakers weren't obvious about it. This is due to a mix of internal regulations and social norms. Before the Motion Picture Association of America enacted its ratings system in 1968, the Hays Code (1934–1968) banned 'sexual perversion' in films, in which homosexuality was certainly considered at that time. Queerness has long been a controversial representation in US films. What results is decades of queer representations that are subtle and clustered around problematic tropes including the male sissy and evil lesbian."

Meg: ***"We've seen that a lot!"***

Melody Lynn Hoffman: "The depictions of queer characters in horror films are much more common to see throughout the last ninety years compared to other film genres. Quantity does not equal quality, though. One of the most pervasive queer depictions in horror is the evil lesbian, most often represented as a vampire. The first time we see this depiction is through the Countess in *Dracula's Daughter* (1936). The Countess is a vampire who focuses her attention and possessiveness on young women. You can see the evil lesbian character spill over to the seemingly innocuous Disney films including 1937's *Snow White and the Seven Dwarfs* (The Evil Queen Grimhilde) and 1989's *The Little Mermaid* (Ursula). This trope relies on an older lesbian character obsessing over a younger woman.

Kelly: ***"That is fascinating!"***

Melody Lynn Hoffman: "As with many queer representations in film, the queerness of the evil woman is often a wink and nod to the audience. Queer audiences have long been hungry to see themselves on screen, so it wasn't hard for them to see the winks and nods. This led to queer audiences identifying with problematic characters. 'Scarcity dictates that sometimes you latch onto some problematic favorites,' Sara Century wrote in her work on queer women in horror. In 2019, there is still a lack of powerful queer characters in horror films. If they aren't the sissy or evil lesbian, they are getting killed off quite early in the films—which is never a good sign. Of course, there are exceptions to this rule, but by and large queerness is still portrayed as taboo in horror films."

Kelly: **"How do you see the future of entertainment in its inclusivity? Are we on the right path?"**

Melody Lynn Hoffman: "Even though the vast majority of horror films have problematic representations of queerness, we are currently experiencing a huge shift in the US film industry that may lead to more diverse representations moving forward. Within the horror genre, we can't ignore the Jordan Peele Effect. The success of his two horror films (*Get Out* and *Us*) has proven to the industry that audiences want to see something different from the status quo. The US film industry is a monolith that is almost impossible to tear down, but Peele's mainstream success will hopefully result in continued funding for more inclusive and complex films."

Kelly: **"That would be amazing!"**

Melody Lynn Hoffman: "Simultaneously, streaming services are becoming a norm. People are able to choose exactly what they want to watch and discover films that are not featured in the US theaters. I have a lot of hope for streaming services and their ability to produce extremely inclusive media. On the whole, I do believe we have a chance to see a huge shift in media representations by turning to streaming programming."

Meg: **"You've written about the ability of fandom to incite social change, have you seen examples of this in horror?"**

Melody Lynn Hoffman: "I believe media fandom can lead to social change, but I am skeptical of it happening regularly when we live as consumers rather than citizens. I can say with certainty that queer audiences directly impact the reception of media texts with 'queer readings' and reimagining what could be within specific story lines. I don't think queer horror fans are much interested in utilizing their fandom for social change. At the very least, I do see queer horror fans lifting up films that offer more complex queer representations. Queer media audiences are important voting blocs, and, when queer representation is overtly involved, they can directly impact a film's success."

The *Buffy the Vampire Slayer* series gave us complex, interesting female characters who were represented in a variety of ways. Women with different personalities, talents, and flaws were portrayed with care showing agency, character growth, and empowerment. The series ends with the notion that we all can be slayers, or do anything that we want, because the power lies within us. We only need to lift one another up and find it within ourselves to fight.

CHAPTER TWENTY-NINE
A QUIET PLACE

Year of Release: 2018	
Director: John Krasinski	
Writer: John Krasinski	
Starring: Emily Blunt, Millicent Simmonds	
Budget: $17 million	
Box Office: $340.9 million	

According to the World Health Organization, over 5 percent of the world's population has hearing loss. It is estimated that by 2050, one in every ten people will have disabling hearing loss. Rosemarie Garland Thomson has written about the paradox that a female body with a disability presents culturally: "Both the female and the disabled body are cast within cultural discourse as deviant and inferior; both are excluded from full participation in public as well as economic life; both are defined in opposition to a valued norm which is assumed to possess natural corporeal superiority."[1]

How are differently abled women portrayed within the horror film genre? A movie that features a deaf female protagonist is *A Quiet Place* (2018). Based on the premise that humans must be silent in order to escape death by creatures who cannot see them, *A Quiet Place* makes for a unique visual and audio experience for the audience.

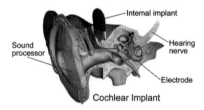

Cochlear implants can help provide a sense of sound to those who are hard of hearing or to profoundly deaf people.

Sounds, or lack of noise, play a major role in the film and give viewers a sense of place and immersion into the world of the characters. Writer and director John Krasinski insisted on casting a deaf actress for the role of his daughter in the film. Millicent Simmonds plays Regan, the congenitally deaf daughter of the couple in the movie. When the audience shifts perspective, and sees things from Regan's point of view, we watch from her silent perspective. On casting Simmonds, Krasinski said, "It was nonnegotiable for me to hire a deaf actress, but I didn't know I was going to get so lucky to have not only a deaf actress, but the most beautiful human being who would walk me through the experience and be honest about what it meant to be deaf."[2]

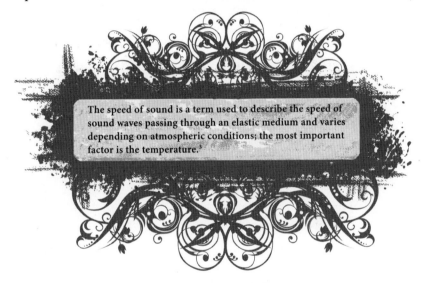

The speed of sound is a term used to describe the speed of sound waves passing through an elastic medium and varies depending on atmospheric conditions; the most important factor is the temperature.[3]

Other horror movies over the years have portrayed women with disabilities, but not all have cast actresses who are differently abled. In *What Ever Happened to Baby Jane?* (1962), Joan Crawford portrays Blanche, a woman who was paralyzed from the waist down in a mysterious car crash. The film explores a dark plot with Jane (Bette Davis) denying her sister food and communication to the outside world. A character being limited by their mobility also plays a part in the plots of *Annabelle: Creation* (2017) and *Silver Bullet* (1985).

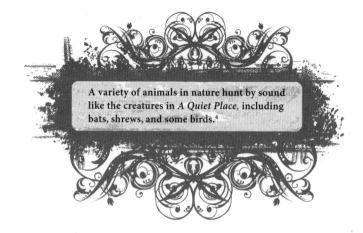

A variety of animals in nature hunt by sound like the creatures in *A Quiet Place,* including bats, shrews, and some birds.[4]

Blindness has been portrayed in horror films like *Wait Until Dark* (1967), *The Eye* (2002), and *Dearest Sister* (2016). Each features a plot where being unable to see comes into play. In the case of *The Eye* and *Dearest Sister,* renewed vision comes with an additional endowment: the ability to see the dead. Although the women playing these roles had sight, there are working actors who are blind. One, Marliee Talkington, has appeared on stage and screen and is one of only two blind people in the United States to have completed an MFA degree in acting. On her experience auditioning, she said:

> I decided I wouldn't tell the directors or anyone on the casting end that I was visually impaired. Which always felt like a betrayal. And I would show up as early as I needed to to rewrite the entire script by hand in large print. I hadn't learned how to vocally advocate for myself yet in a way that didn't feel angry or demanding, so at times I flat out lied.[5]

Although Talkington has been getting work, a study found that "more than 95 percent of characters with disabilities are played by able-bodied actors on television. While streaming platforms had a better percentage, they also had a lower overall count of characters with disabilities. This lack of self-representation points to a systemic problem of ableism—discrimination against people with disabilities—in the television industry. It also points to a pervasive stigma among audience members against people with disabilities given that there is no widespread outcry against this practice."[6]

Hush (2016) and *The Shape of Water* (2017) both feature female leads with deafness, although each was portrayed by a hearing actress. Kate Siegel, who portrays Maddie in *Hush* cowrote the movie with her husband and director, Mike Flanagan. Siegel acknowledged her regret on not casting a deaf actress in the role:

> I learned that I was being offensive, and that was very difficult for me. . . . Something that people liked about *Hush* was that her weakness became her strength. I learned a lot about hearing privilege, white privilege, and about the assumptions I made before I wrote the script. I'm grateful for the knowledge I got about it; I learned a lot about my privilege as a hearing actress."[7]

Hush is the story of a deaf writer who lives in a remote house in the woods. She is plagued by a masked assailant who ends up murdering her sister and her boyfriend. A unique twist to the plot is that Maddie can "see" various endings to her story, just like she does when writing, so she can decide how best to escape. On writing the story, Siegel said:

> Nothing happens to Maddie because she's a woman and she doesn't choose anything because she's a woman. We could neutralize gender in this movie and you would have the exact same movie. That to me was very important and I wanted to make sure that my female friendship with Samantha [Sloyan, who plays Sarah in the film] wasn't about boys. It was about reading a book and talking about books. I wanted to make sure the relationship between the sisters was familial and didn't need to be girly and giggly. I had a strong eye on that most of the time. A lot of this movie is, with a risk of putting too much into it, a metaphor for feeling unheard. It's a movie about asserting yourself and of course as a female writer I brought a lot to that.[8]

Maddie is portrayed as a strong, capable woman who defeats her would-be killer at a remote house in the woods. But not all people with disabilities have been portrayed this way. Some Puritan writings suggested that disabled people were:

> ... innately driven toward evil, and that a child born with a disability was being punished for intrinsic impulses towards immorality. In folktales and fairytales, too, limping crones lure children to their deaths, and disfigured characters like Rumpelstiltskin use their cruelty and cunning to entrap the more moral characters of the story. Often, in these tales, disabled and disfigured people are fueled by jealousy and bitterness, and so turn their hatred onto the pure and blameless members of society. The overarching message? Disabled people are inherently evil, and as such, we are scary.[9]

According to a study by The Annenberg Foundation, only 2.4 percent of all speaking or named characters in film were shown with a disability.[10] Hopefully, with more initiatives and education, representation in horror, and all genres, can be improved. One way this can happen is to have writers and directors insist on casting actors who possess the traits of the characters in their films. Regan, in *A Quiet Place*, is portrayed with realism and nuance that any other performer may have missed. The character is allowed to show quite a range of emotion in the film including joy, grief, guilt, fear, and anger. She is shown experiencing weakness but ultimately becomes the hero of her own story by defeating the creatures plaguing her family. Her hearing loss becomes her strength instead of her downfall, and the audience feels it with her. Accurate representation leads to great empathy, which can only lead to a more fulfilling filmgoing experience.

CHAPTER THIRTY
KICK-ASS WOMEN OF THE PAST AND THE FUTURE

Women have been involved in the genre of horror since its inception and will continue to be creators, designers, and performers. Who are some of the earliest examples of women in the horror film world? Alice Guy-Blaché shot hundreds of films between 1896 and 1920, including several horror movies: an adaptation of Edgar Allan Poe's *The Pit and the Pendulum* (1913), *The Monster and the Girl* (1914), and *The Vampire* (1915). Her films used many cutting-edge techniques, and her filmmaking directly influenced Alfred Hitchcock, among others. "I was thrilled by the movies of D.W. Griffith and the early French director Alice Guy," he once told an interviewer.[1]

Lois Weber is another notable early pioneer in the horror genre. Weber wrote, directed, and starred in a silent film in 1913 titled *Suspense*, which featured an uncredited role by soon-to-be horror legend Lon Chaney. The thriller used the split-screen technique, which Weber pioneered, and included a car chase. She is also credited with being the first American woman to direct a feature-length film and the first to own her own film studio.

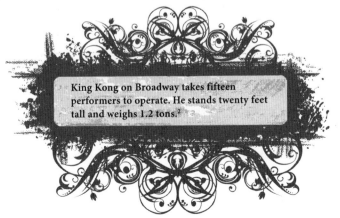

King Kong on Broadway takes fifteen performers to operate. He stands twenty feet tall and weighs 1.2 tons.[2]

Ruth Rose is the screenwriter who made *King Kong* (1933) into the fast-paced, thrilling adventure it was. Rose created a monster who was sympathetic and characters who allowed the audience to access their empathy for them. Previous to Rose's involvement, James Ashmore Creelman had crafted a much slower vision of *King Kong*, and it was Rose who gave Carl Denham and Jack Driscoll their personalities, which guided much of the energy of the picture. "Oh, no. It wasn't the airplanes. It was Beauty killed the Beast," perhaps *King Kong's* most famous line, was written by Rose.[3] In 2018, King Kong premiered as a larger-than-life character on Broadway that emulated heart and personality eight shows a week.

Ruth Rose wrote King Kong to be a sympathetic character.

Another woman who created an iconic monster for horror cinema was Milicent Patrick. She was one of the first female animators at Walt Disney Studios and later became a special effects makeup artist. Patrick most famously created the Gill-man from *Creature From the Black Lagoon* (1954), which inspired generations of filmmakers and designers in the process. One notable director, Guillermo del Toro, said, "The creature was the most beautiful design I'd ever seen and I saw him swimming under [actress] Julie Adams, and I loved that the creature was in love with her, and I felt an almost existential desire for them to end up together. Of course, it didn't happen."[4] Ultimately, he created his own love story for the two in 2017's *The Shape of Water* and drew inspiration from the movie that intrigued him as a child, calling the creature the "number one monster suit in film history."[5]

Daphne du Maurier is a writer who contributed to the horror literature genre and also had her stories turned into successful films. Du Maurier wrote *Rebecca*, which was turned into a film in 1940, and *The Birds*, which became the Alfred Hitchcock classic in 1963. A more recent adaptation of her work is *My Cousin Rachel* (2017), but her past work remains timeless. Du Maurier's type of horror is profound and unsettling because her monsters are real people, out in the daylight. We can imagine the situations she puts her characters in because they seem plausible and realistic.

Ricou Browning, a professional diver and swimmer, was required to hold his breath for up to four minutes at a time for his underwater role as the "Gill-man."[6]

Filmmaker Ida Lupino is considered to be the first female film noir director in the United States for *The Hitchhiker* (1953) and the first female director of an episode of *The Twilight Zone* (1958–1964). She directed the episode titled "The Masks" in 1964, in which a man forces his relatives to wear masks that reflect their true personalities in order to receive their inheritance from him. Their faces transform to represent their true selves and become exactly like the hideous masks.

Another notable woman who directed iconic horror movies is Mary Lambert, who directed *Pet Sematary* (1989) and *Pet Sematary Two* (1992). She was the first woman to be in charge of a major studio horror film, and *Pet Sematary* spent three weeks at the number one spot at the box office, at the time the most successful movie helmed by a female director. Lambert gained Stephen King's approval and recalled:

I wasn't looking to direct a horror film. I liked Stephen King though, and I had read almost all of his books. Nobody could read all of his books at this point in time, unless that's all they did, because he's so prolific. I had read a number of his novels, including *Pet Sematary*, and I got really excited about it and realized it would be a great progression for me.[7]

We spoke with Vanessa Ionta Wright, the director of the Women in Horror Film Festival (WIHFF), to learn more about the festival itself and her experience working in the genre.

Kelly: **"What made you start the Women in Horror Film Festival?"**

Vanessa Ionta Wright: "To put it quite simply, there was a need for this festival. It's no secret that there is a huge gender gap in film, especially in the horror genre. According to a recent study, of the two hundred and fifty top box office grossing films in 2018, only 8 percent were directed by women, which is 3 percent less than 2017. The Women in Horror Film Festival was launched to put a spotlight on women in independent genre film, and to help shrink that gender disparity. Our goal is to showcase women beyond the screen and even beyond the director's chair. We are proud to recognize the contributions of female producers, cinematographers, writers, composers, editors, production designers, and SFX artists in film. The mission of the fest is to promote inclusion and equality and that is exactly what we look for in a film that screens at WIHFF."

Kelly: **"What has surprised you the most during these past two years that the festival has been running?"**

Vanessa Ionta Wright: "As far as what has surprised me the most, is how much support the fest has received in its first two years. When Samantha and I launched the fest in 2017, we were not expecting it to take off as quickly as it did. We had a 49 percent increase in submissions from year one to year two. I'm thrilled that people are responding so positively to the fest and the mission behind it, and I will definitely continue to nurture this event and bring to audiences the best in independent horror cinema from around the world."

Meg: **"Tell us about your experience directing *Rainy Season* (2017) and *I Baked Him a Cake* (2016). What drew you to these stories?"**

Vanessa Ionta Wright: "My experience directing *Rainy Season* was simultaneously terrifying and exhilarating. I had taken a nearly ten-year break from film. I started writing again in 2014 and received a couple of accolades for my work and decided to take the next logical step for myself as a filmmaker and direct a film. I had heard through the grapevine that Stephen King will release certain short stories for the low price of $1 to aspiring filmmakers through his Dollar Baby program. I read through the available list and recognized many of the stories, but I hadn't heard of "Rainy Season." I grabbed my copy of *Nightmares & Dreamscapes* (1993) and immediately fell in love and decided that this needed to be my first film. I applied for the rights, signed the contract, and mailed Mr. King $1. As far as what drew me to the story, I think it was the desperation and what the citizens of Willow had to do in order for their town and their lives to prosper. It reminded me so much of Shirley Jackson's "The Lottery" (1948), which King actually references just a few pages in. I felt a strange connection to this story, so I gathered an extremely talented cast and crew and we made it happen. We were a month away from shooting *Rainy Season* when I traveled to Pennsylvania to direct Samantha's script, *I Baked Him a Cake.* Samantha had crafted a very contained and very disturbing story that occurs in a five-minute run time. What a wonderful challenge to tell a story and evoke a reaction from the audience in such a small time frame! What drew me to *I Baked Him a Cake* was similar to what drew me to *Rainy Season,* an overwhelming sense of dread and desperation. Both stories leave you with a sense of 'Oh, shit, what happens next?' I love the satisfaction that comes with wanting more, and I think audiences walk away from both films with that desire for more."

Meg: **"As a woman do you find your creativity piqued by female-driven content?"**

Vanessa Ionta Wright: "Sure, if the content is compelling. Are the characters complex? Does the story excite me? Does it stick with me long after I've watched/read it? Does it offer a fresh perspective? These are the things that I look for in a script, and none of it has anything to do with gender . . . it's all about the story."

Kelly: **"Do you remember seeing your first horror movie? What was it, and how did it impact your life?"**

Vanessa Ionta Wright: "I guess my love of watching horror movies has impacted my life greatly, considering I make horror films and run a horror film festival. The first horror movie that I remember watching start to finish was *Poltergeist* (1982). It scared the shit out of me! I wasn't supposed to see it. That was probably the driving force behind my decision to watch it. Ye olde forbidden fruit. I was eight years old and living just outside of Boston, Massachusetts. A friend of mine invited me over and we watched it on her TV and that was it. I was terrified. So, what did I do? I went looking for more scares, of course. As frightened as I was, I was also so drawn to horror. Halloween was (and still is) my favorite holiday, but I was scared of everyone in a costume. Each trick-or-treat night generally ended with me hiding under my bed until it was over. Then, I would start planning my costume and trick-or-treat route for the next year. It was the same way with movies. They would scare the ever-living crap out of me and I would protest to my parents about my bedtime and having to go upstairs alone, but then the next day I was searching for another horror movie to watch. I'm the same way with haunted houses that pop up every October, I don't dare step foot in them, I'm too scared. To this day, at forty-three years old, I will finish a scary movie, turn off the lights downstairs, and race up the stairs in the dark like I'm being chased by something. I'm probably being chased by something."

Kelly: **"Every night when I walk down my dark hallway to the bedroom, I picture Samara from *The Ring* lurking around the corner!"**

Meg: **"The kick-ass woman is a popular archetype in horror. Is this a particular character trope you relate to and seek out as a creator and audience member?"**

Vanessa Ionta Wright: "On a conscious level, I don't think I seek it out, but I'm thrilled to see more films showcasing female characters that are strong and intelligent. These are the characters that I think we all can, and want to, relate to. Who wants to be victim? Ask anyone who they would want to be, and more than likely they will reply with 'the hero'; it's empowering. For far too long, women have been portrayed in film and literature as submissive, vulnerable, and weak, especially in horror: the damsel in distress who needs a man to rescue her. The bigger the boobs, the lower the IQ and the more likely that they'll be killed off before we get to the second reel. It's insulting, it's always been insulting. It's nice to see a character fully realized in her own power, acknowledging that she can not only save herself, but everyone else around her. We've had these stereotypes shoved down our throats that the pretty, buxom girl is stupid and promiscuous and will probably trip over her boyfriend's dead body or a poorly placed bong, because we all know if you have sex or do drugs you're gonna die. But not to worry, because your plain-faced, flat-chested, bookworm of a friend, who, by the way, is still a virgin, will save the day . . . but she won't get the guy at the end, because he's dead, along with all of her friends and probably her parents. I'm sorry, I think my eyes just rolled completely to the back of my head for a second and I'm rambling. Bottom line, I think it's amazing that more roles and story lines are being written outside of these stereotypes and tired tropes. I think it's great that many more stories are being put into the hands of female visionaries who are creating amazing films. Women can own their sexuality, they can exhibit strength and intelligence, and they can face whatever monster or adversity is thrown at them and win, both on and off the screen."

Kelly: **"What do you see as the future for women in the horror genre?"**

Vanessa Ionta Wright: "Now, and in the future, I see more opportunities. More important, I see women creating their own opportunities and taking ownership of the films they want to create and the stories they want to tell. I think we will eventually see more balance and diversity both on-screen and behind the camera. It's a long road, filled with bumps and potholes, but a road with the destination clearly in our sights, and we have a full tank of gas."

Meg: "**Tell us about your current projects and where we can see your work next!**"

Vanessa Ionta Wright: "I'm working on directing a couple of short scripts that I have written. One is a very dark thriller called *Invidia,* and the other is an over-the-top horror comedy called *Permanent Damage.* I've also recently been hired to write a feature, a reimagining of a 1973 film, and I've put my name in the hat to direct it, as well. A screenwriter friend of mine, Edward Santiago, recently asked me to direct his short western script, which I'm really excited about. I stay quite busy working on WIHFF, and 2020 is bringing a lot of exciting changes with the fest. I'm easy to keep up with, I stay pretty active on Facebook, Instagram, and Twitter, so be sure to follow along on all my adventures in film."

The Women in Horror Film Festival is just one of the many ways to celebrate women in the genre of horror in the month of February or all year round. There are numerous women making movies currently in the genre of horror, and the website Cut-Throat Women provides a database of directors, producers, screenwriters, and film festivals in the field.

The Soska Sisters, Jen and Sylvia, have written and directed numerous horror films together including *American Mary* (2012) and *Rabid* (2020), a remake of the David Cronenberg movie by the same name. Jen Soska said, "People ask us, 'Why do you do this? Why do you speak out so much?' It's because I want it to be easier for other women. Because there are so many girls who come to us and say they want to be directors . . . it feels wonderful in a way to be trailblazing so that maybe, in a way, if we go through this, we set a precedent . . . it's going to be easier for other girls down the line."[8]

Julia Ducournau is another director who has made her mark in the horror world. Her 2016 film *Raw* tells the story of a vegetarian college student in her first week of veterinary school who develops a taste for human flesh. The movie is also a coming-of-age story and explores the main characters' sexuality. Ducournau said:

> I'm fed up with the way young women and their discovery of sexuality is portrayed on screens. I feel it's always a victim's story that's being told. It's always about the fear, or the doubt afterwards . . . For me, sexuality is in the body. And you should certainly not be a victim. It's not something that you go through, it's something that you are active in, and it's perfectly okay.[9]

When Mary Shelley wrote *Frankenstein,* she couldn't have predicted the ripple effect it would have for both horror and women. As two girls who read her work, we became entranced with the macabre, as well as the notion that we could become creators of what we loved. Now, as women, we feel we're living in a horror renaissance, a time at which females are controlling their narratives through the vivid and exciting world of horror. Through the lens of science, we can better understand both the limitations and the vastness of horror and its interplay with gender. As Julia Ducournau asserts, women should take agency of their narratives in the horror genre. It's time for women to tell their own stories and own them, without fear. As Mary Shelley wrote, "I am fearless and therefore powerful."

ACKNOWLEDGMENTS

Thank you to all the women who took part in interviews for this book: Alice L., Dee, Jenna, Ashlee, Stacey, Alice S., Jen, Sylvia, Deborah, Aislinn, Shayna, Sam, Gigi, Melody, and Vanessa.

Thank you to Nicole Mele and everyone at Skyhorse Publishing.

Love and thank you to Luke, Mark, and Lisa for the help and support. And to our Rewinders, we'll see you in the horror section!

ABOUT THE AUTHORS

Kelly Florence (left) and Meg Hafdahl

Kelly Florence is a communication instructor at Lake Superior College in Duluth, Minnesota, and is the creator and cohost of the *Horror Rewind* podcast as well as the producer and host of the podcast *Be A Better Communicator*. She received her BA in theater at the University of Minnesota-Duluth and earned her MA in communicating arts at the University of Wisconsin-Superior. Kelly is the coauthor of *The Science of Monsters* also from Skyhorse Publishing.

Horror and suspense author Meg Hafdahl is the creator of numerous stories and books. Her fiction has appeared in anthologies such as *Eve's Requiem: Tales of Women, Mystery, and Horror* and *Eclectically Criminal*. Her work has been produced for audio by *The Wicked Library* and *The Lift*, and she is the author of two popular short story collections including *Twisted Reveries: Thirteen Tales of the Macabre*. Meg is also the author of the two novels *Daughters of Darkness* and *Her Dark Inheritance*, called "an intricate tale of betrayal, murder, and small-town intrigue" by *Horror Addicts* and "every bit as page turning as any King novel" by *RW* magazine. Meg, also the cohost of the podcast *Horror Rewind* and coauthor of *The Science of Monsters*, lives in the snowy bluffs of Minnesota.

ENDNOTES

CHAPTER ONE: PREVENGE

1. North, Anna. (April 23, 2019) "You're Doing It Wrong: The Century-Old Roots of Mom-Shaming." *Vox.*
2. Lowe, Alice. (2017) "The Prevenge Diaries." *Raising Films.*
3. (March 5, 1945) "Medicine: Prodigious Pregnancy." *Time Magazine.*
4. Erbland, Kate. (March 23, 2017) "How a Pregnant Actress Overcame Her Frustrations with the Industry and Directed Her First Feature." *Indie Wire.*
5. (February 14, 2017) "Alice Lowe: 'It Wasn't Part of the Plan to Direct while Pregnant.'" *The Guardian.*
6. Lowe, Alice. "The Prevenge Diaries." *Raising Films.*
7. Luzen, Math M., PhD (2019) "The Celluloid Ceiling: Behind-the-Scenes Employment of Women on the Top 100, 250, and 500 movies of 2018." *Center For the Study of Women in Television and Film.*
8. Levine, Jeff. (March 20, 2001) "No. 1 Cause of Death in Pregnant Women: Murder." *WebMD.*
9. Good Therapy Staff. (May 27, 2018) "Grieving Through Pregnancy? What You Should Know." *Good Therapy.*
10. Hytten, F. (1985) "Blood Volume Changes in Normal Pregnancy." *Clin Haematol.*
11. Chatterjee, Rhitu. (March 21, 2019) "New Postpartum Depression Drug Could Be Hard to Access for Moms Most in Need." *NPR.*
12. (2019) "Postpartum Psychosis." *Postpartum Support International.*

CHAPTER TWO: THE BABADOOK

1. Adams, Sam. (November 27, 2014) "Boogeyman Nights: The Story Behind This Year's Horror Hit The Babadook." *Rolling Stone.*
2. Brogaard, Berit. (November 9, 2016) "Parental Attachment Problems." *Psychology Today.*
3. Lambie, Ryan. (October 13, 2014) "Jennifer Kent Interview: Directing The Babadook." *Den of Geek.*
4. Enoch, M. David. (2001) *Uncommon Psychiatric Syndromes.* CRC Press.
5. Parker, Laura. (December 5, 2014) "A Woman Directed the Scariest Movie of the Year, Maybe the Decade." *The Cut.*
6. Enoch, M. David. (2001) *Uncommon Psychiatric Syndromes.* CRC Press.

7. (November 14, 2014) "Parental Descent: Jennifer Kent's The Babadook Is a Spooky of a Mother in Crisis." *Film Journal International.*

8. Havrilesky, Heather. (November 8, 2014) "Our Mommy Problem." The *New York Times.*

9. Barone, Matt. (November 28, 2014) "The Year's Best Horror Movie? It's This Australian Creepshow, Hands Down." *Complex.*

10. Barone, Matt. (November 28, 2014) "The Year's Best Horror Movie? It's This Australian Creepshow, Hands Down." *Complex.*

11. Adams, Sam. (November 27, 2014) "Boogeyman Nights: The Story Behind This Year's Horror Hit The Babadook." *Rolling Stone.*

12. Lambie, Ryan. (October 13, 2014) "Jennifer Kent Interview: Directing The Babadook." *Den of Geek.*

13. Gress, Jon. (2015) *Visual Effects and Compositing.* New Riders.

CHAPTER THREE: BATES MOTEL

1. Chocano, Carina. (April 24, 2017) "How Do You Reclaim the Mother from Psycho?" *The Cut.*

2. Ibid.

3. (January 9, 2015) "The Long, Strange, 60-Year Trip of Elmer McCurdy." *NPR.*

4. Johnson, Madison Alisa. (May 2016) "The Woman As Place: The Utilization of the Female Body in Horror Film." *Clemson University.*

5. VanDerWerff, Emily. (May 10, 2016) "Bates Motel's Norma Bates Is TV's Best and Worst Mother." *Vox.*

CHAPTER FOUR: A NIGHTMARE ON ELM STREET

1. Clover, Carol. (1992) *Men, Women, and Chainsaw: Gender in the Modern Horror Film.* Princeton University Press.

2. Eaton, E.W. (September 2008) "Feminist Philosophy of Art." *Philosophy Compass.*

3. Totaro, Donato. (January 2002) "The final girl: A Few Thoughts on Feminism and Horror." *Off Screen.*

4. Muir, John. (2012) *Horror Films of the 1980s,* Volume 1. McFarland.

5. Borelli, Lizette. (March 31, 2015) "A Bad Dream is More Than Just a Bad Dream: The Science of Nightmares." *Medical Daily.*

6. Collis, Clark. (July 29, 2011) "Nightmare on Elm Street: Whatever Happened to Nancy?" *Entertainment Weekly.*

7. Cayer, Ariel Esteban. (June 1, 2011) "I Am Nancy—An Interview with Heather Langenkamp." *Spectacular Optical.*

8. Cupp, Lauren. (2017) "The final girl Grown Up: Representations of Women in Horror Films From 1978–2016." *Scripps College.*

9. Dickson, Evan. (January 18, 2013) "10 Facts about the New Evil Dead!" *Bloody Disgusting.*

CHAPTER FIVE: TALES FROM THE CRYPT: DEMON KNIGHT

1. Rahbaran, Saeed. (August 31, 2018) "What It's Like to Be a Black Female Stunt Performer." *The Undefeated.*
2. Wixson, Heather. (January 12, 2015) "Retrospective: Celebrating the 20th Anniversary of Tales From the Crypt: Demon Knight. Part One." *Daily Dead.*
3. Miska, Brad. (June 25, 2018) "Joel Silver Wanted Cameron Diaz to Star in Tales From the Crypt: Demon Knight." *Bloody Disgusting.*
4. Leeuwen, Jacoba. (2006) *Symbolic Communication in Late Medieval Towns.* Leuven University Press.
5. Benshoff, Harry M. (2000) "Blaxploitation Horror Films: Generic Reappropriation or Reinscription?" *Cinema Journal.*
6. Wixson, Heather. (January 13, 2015) "Retrospective: Celebrating the 20th Anniversary of Tales From the Crypt: Demon Knight. Part Two." *Daily Dead.*
7. Wise, Kathy. (October 2009) "Georgina Lighting: the First Native Female Director of a Feature-Length Film." *Cowboys & Indians.*
8. Romero, Ariana. (April 26, 2019) "Chambers' Sivan Alyra Rose Grew Up on a Rez—Now She's Changing Netflix Forever." *Refinery 29.*

CHAPTER SIX: HALLOWEEN

1. Hur, Johnson. (November 26, 2016) "The History of Babysitting." bebusinessed.com.
2. Berry, Craig. (March 30, 2018) "The Unsolved Murder of Janett Christman." truecrimearticles.com.
3. Kaminski, Julia. (February 10, 2017) "Ode to Debra Hill." *Scream Fest.*
4. (November 18, 2009) "Halloween." *History.com.*
5. Kay, Don. (October 17, 2018) "Halloween: Jamie Lee Curtis on the Trauma of Laurie Strode." *Den of Geek.*
6. (July 6, 2018) "Post-Traumatic Stress Disorder." mayoclinic.org.
7. Taylor-Foster, Kim. (October 25, 2018) "Halloween Flips the final girl Trope On its Head." *Fandom.*
8. Bretherton, I. (1990) "Communication Patterns, Internal Working Models, and the Intergenerational Transmission of Attachment Relationships." *Infant Mental Health Journal*, 11(3), 237–251.
9. McGuire, Virginia Claire. (April 4, 2013) "10 Real-Life Panic Rooms." *Mental Floss.*

CHAPTER SEVEN: SLEEPAWAY CAMP

1. Galupo, M. Paz. (June 2014) "Sexual Minority Reflections on the Kinsey Scale and the Klein Sexual Orientation Grid: Conceptualization and Measurement." *Journal of Bisexuality.*
2. Maclay, Willow. (August 10, 2015) "How Can It Be? She's a Boy. Transmisogyny in Sleepaway Camp." *Cleojournal.com.*
3. Brehmer, Nat. (September 3, 2017) "The Pros and Cons of Sleepaway Camp as a Trans Narrative." *Wickedhorror.com.*
4. Serena, Katie. (January 9, 2019) "The Lake Bodom Murders: Finland's Most Famous Unsolved Triple Homicide." *All That's Interesting.*
5. Zimmerman, Bonnie. (March 1981) "Daughters of Darkness Lesbian Vampire Films." *Jump Cut.*
6. Bevan, Thomas E. (2014) *The Psychobiology of Transsexualism and Transgenderism.* Praeger.

CHAPTER EIGHT: TEETH

1. Molitor, Fred, Sapolsky, Barry S. (Spring 1993) "Sex, Violence, and Victimization in Slasher Films." *Journal of Broadcasting & Electronic Media.*
2. Welsh, Andrew. (2010) "On the Perils of Living Dangerously in the Slasher Film: Gender Differences in the Association Between Sexual Activity and Survival." *Sex Roles.*
3. Rieser, Klaus. (April 2001) "Masculinity and Monstrosity: Characterization and Identification in the Slasher Film." *Austria Men and Masculinities.*
4. Koehler, Sezin.(June 15, 2017) "Pussy Bites Back: Vagina Dentata Myths From Around the World." *Vice.*
5. Valenti, Jessica. (2009) *The Purity Myth: How America's Obsession with Virginity Is Hurting Young Women.* Seal Press.
6. Pheterson, Gail. (Winter 1993) "The Whore Stigma: Female Dishonor and Male Unworthiness." *Social Text.*
7. Renner, Karen J. (Spring 2016) "Monstrous Schoolgirls: Casual Sex in the Twenty-First-Century Horror Film." *Northern Arizona University Red Feather Journal.*
8. Morrow, Brendan. (April 27 2016) "'It Follows' is Not About STDs. It's About Life as a Sexual Assault Survivor." *Bloody Disgusting.*
9. Adams, Letizia. (October 20, 2017) "Only 13 States Require Sex Ed to Be Medically Accurate." *Vice.*

CHAPTER NINE: GERALD'S GAME

1. Kalush, William. (2007) *The Secret Life of Houdini: The Making of America's First Superhero.* Atria Books.

2. Black, M.C., et al. (2011) "The National Intimate Partner and Sexual Violence Survey (NISVS): 2010 Summary Report." *National Sexual Violence Resource Center.*

3. Rao, Joe. (July 1, 2019) "Solar Eclipses: When Is the Next One?" *Space.com.*

4. Wood, Robin. (1987) "Returning the Look: Eyes of a Stranger." *American Horrors: Essays on the Modern American Horror Film.*

5. Rieser, Klaus. (April 2001) "Masculinity and Monstrosity: Characterization and Identification in the Slasher Film." *Austria Men and Masculinities.*

6. Hirschon, Renée. (1978) "Open Body/Closed Space: The Transformation of Female Sexuality." *Defining Females: the Nature of Women in Society.*

CHAPTER TEN: US

1. Cera, Diego. (February 23, 2017) "7 Paintings of Betrayal and Revenge to Understand Humanity's Most Despicable Side." *Cultura Colectiva.*

2. Jaffe, Eric. (October 2011) "The Complicated Psychology of Revenge." *Association for Psychological Science.*

3. McKee, Ian, PhD. (June 2008) "Revenge, Retribution, and Values: Social Attitudes and Punitive Sentencing." *Social Justice Research.*

4. Price, Michael. (June 2009) "Revenge and the People Who Seek It." *American Psychological Association.* Volume 40, No. 6.

5. Gutowitz, Jill. (January 19, 2018) "I'm Tired of Male Screenwriters Using Rape as a Convenient Backstory For Women." *Glamour.*

6. Damas Mora, J.M., Jenner, F.A., Eacott, S.E. (1980) "On Heautoscopy or the Phenomenon of the Double: Case Presentation and Review of the Literature." *Br J Med Psychol.*

7. Brugger, P. (2002) "Reflective Mirrors: Perspective-taking in Autoscopic Phenomena." *Cognitive Neuropsychiatry.*

8. Wells, H.G. (1895) *The Time Machine.* William Heinemann, United Kingdom.

9. Brantley, Carol. (2003) "Class: Power, Privilege, and Influence in the United States." *Workforce Diversity Network.*

10. Ibid.

11. Ling, Lisa. (September 23, 2009) "Under Las Vegas: Tunnels Stretch for Miles." *ABC News.*

12. Bushman, Brad J. (2002) "Does Venting Anger Feed or Extinguish the Flame? Catharsis, Rumination, Distraction, Anger, and Aggressive Responding." *Iowa State University.*

13. McCullough, Michael E. (2008) *Beyond Revenge: The Evolution of the Forgiveness Instinct.* Jossey-Bass.

CHAPTER ELEVEN: WHAT LIES BENEATH

1. Ramm, Michaela. (October 28, 2014) "Statistics for You: Haunted Houses." *Iowa State Daily*.
2. Crooke, William. (1894) *An Introduction to the Popular Religion and Folklore of Northern India*. Forgotten Books.
3. Ramm, Michaela. (October 28, 2014) "Statistics for You: Haunted Houses." *Iowa State Daily*.
4. Grossman, Pam. (May 17, 2000) "Girls School Rules." *Salon*.

CHAPTER TWELVE: A GIRL WALKS HOME ALONE AT NIGHT

1. Anyiwo, Melissa U. (2016) "The Female Vampire in Popular Culture." *Teaching Gender*.
2. Ghigi, Giulia. (2012) "Interview to Xan Cassavetes, Director of Kiss of the Damned." *Venice Film Festival*.
3. Salovaara, Sarah. (January 19, 2014) "Five Questions With A Girl Who Walks Home Alone At Night Director Ana Lily Amirpour." *Filmmaker Magazine*.
4. Mayo Clinic Staff. (November 18, 2017) "Pohphyria: Symptoms and Causes." Mayoclinic.org.
5. Miranda, Carolina A. (January 12, 2015) "Director Ana Lily Amirpour on Iranian Vampires and Weird SoCal towns." *Los Angeles Times*.
6. O'Hehir, Andrew. (November 21, 2014) "A Girl Walks Home Alone At Night: The Black-and-White, Feminist Iranian Vampire Western You've Been Waiting For." *Salon*.
7. Balswick, J., & Peck, C. (1971) "The Inexpressive Male: A Tragedy of American Society." *The Family Coordinator*. 20, 363–368.
8. Cakir, Deniz. (February 25, 2018) "A Girl Walks Home Alone At Night Is the Feminist Movie of Our Dreams, Literally." *Medium*.
9. Mandal, Dattatreya. (July 1, 2019) "Scythians: 12 Things You Should Know about the Ancient Horselords of The Steppe." *Realm of History*.
10. Machiavelli, Niccolo. (1532) *The Prince*. Antonio Blado d'Asola.
11. King, Stephen. (2000) *Riding the Bullet*. Simon and Schuster.

CHAPTER THIRTEEN: LET THE RIGHT ONE IN

1. Burks, Raychelle. (June 16, 2017) "The Acid (Bath) Test." *Chemistry World*.
2. Blair, Robbie. (October 23, 2013) "The Uncanny Factor: Why Little Girls Scare the Shit Out of Us." *Lit Reactor*.
3. Kolb, Leigh. (October 24, 2012) "The Terror of Little Girls: Social Anxiety About Women in Horrifying Girlhood." btchflcks.com.
4. Ericsson, Kjersti, and Jon, Nina. (December 1, 2006) "Gendered Social Control: 'a Virtuous Girl' and 'a Proper Boy.'" *Journal of Scandinavian Studies in Criminology & Crime Prevention*.

5. Hellman, Roxanne. (2011) *Vampire Legends and Myths*. Rosen Pub Group.

CHAPTER FOURTEEN: BEETLEJUICE

1. Shuker, Karl. (2002) *The Unexplained: An Illustrated Guide to the World's Paranormal Mysteries*. Metro Books.
2. Akhouri, Kehksha, and Akhouri, Deoshree (January 1, 2018) "Impact of Parent-Child Relationship on Educational Aspiration and Self-Esteem of Adolescents Boys and Girls." *Uttar Pradesh Indian Journal of Health and Well-being*.
3. Maddalena, Ronald J. (1998) "Betelguese." www.gb.nrao.edu.
4. Markovitz, Adam. (April 11, 2014) "The Original Mean Girls." *Entertainment Weekly*.
5. Ward, Jennifer. (December 2, 2012) "Six Archetypal Horror Characters and Why They're Important." *The Artifice*.
6. Molitor, Fred, and Sapolosky, Barry S. (Spring 1993) "Sex, Violence, and Victimization in Slasher Films." *Journal of Broadcasting & Electronic Media*.

CHAPTER FIFTEEN: WHAT EVER HAPPENED TO BABY JANE?

1. McCulloch, Sara Black. (August 10, 2015) "Girl Fight: Bette Davis and Joan Crawford in What Ever Happened to Baby Jane?" *Cleojournal.com*
2. Gallagher, Caitlin. (March 5, 2017) "Bette Davis' Baby Jane Costume in Feud May Have Been Channeling More Than Her Character." *Bustle*.
3. Boyle, Karen. (2001) "What's Natural About Killing? Gender, Copycat Violence and *Natural Born Killers*." *Journal of Gender Studies*.
4. Mersky Leder, Jane. (January 1, 1993) "Adult Sibling Rivalry." *Psychology Today*.

CHAPTER SIXTEEN: DRAG ME TO HELL

1. Hastings, Christobel. (April 9, 2018) "The Timeless Myth of Medusa, a Rape Victim Turned Into a Monster." *Broadly*.
2. Johnston, Elizabeth. (November 6, 2016) "The Original Nasty Woman." *The Atlantic*.
3. Vasil, Latika, and Wass, Hannelore. (January 1993) "Portrayal of the Elderly in the Media." *Educational Gerontology*.
4. Lauzen, Dr. Martha M. (September 2018) "Boxed in 2017–18: Women on Screen and Behind the Scenes in Television." *Center for the Study of Women in Television & FIlm*. San Diego State University.
5. Thehorrorchick. (June 2, 2009) "Lorna Raver Talks Drag Me to Hell." *Dread Central*.
6. Ibid.

7. West, David R. (1995) Some Cults of Greek Goddesses and Female Daemons of Oriental Origin. Butzon & Bercker.
8. (May 19, 2009) "Production Notes." dragmetohell.net.
9. Winters, Riley. (June 20, 2019) "The Real History of the Romani People and the Misnomer of Gypsies." *Ancient Origins*.
10. Shukla, Ankita. (December 21, 2016) "Depiction of Women in Literature Through the Ages." *Times of India*.

CHAPTER SEVENTEEN: THE VISIT

1. Zwecker, Bill. (September 9, 2015) "Chicago Actress Deanna Dunagan Plays Scary Grandma in The Visit." *Chicago Sun Times*.
2. Stamberg, Susan. (February 26, 2016) "Directors Know: When Child Actors Are on Set, the Studio Teacher is in Charge." *NPR*.
3. Lin, Judy. (January 7, 2010) "Honor or Abandon: Societies' Treatment of Elderly Intrigues Scholar." *UCLA Newsroom*.
4. DeLong, William. (March 1, 2018) "Nannie Doss Spent Decades Murdering Relatives and Husbands." *All That's Interesting*.
5. Willerslev, Rane. (2009) "The Optimal Sacrifice: A Study of Voluntary Death Among the Siberian Chukchi." *American Ethnologist*.
6. Diamond, Jared. (2012) *The World Until Yesterday: What Can We Learn From Traditional Societies?* Viking Press.
7. Aggliass, Kyle, PhD. (September 8, 2014) "Family Estrangement: Aberration or Common Occurrence?" *Psychology Today*.
8. Nanda, Dr. Silima. (2014) "The Portrayal of Women in the Fairy Tales." *The International Journal of Social Sciences and Humanities Invention*. Volume 1, Issue 4.
9. Blair, Elizabeth. (October 28, 2015) "Why Are Old Woman Often the Face of Evil in Fairy Tales and Folklore?" *NPR*.
10. Clifford, Marissa. (November 3, 2017) "The Enduring Allure of Baba Yaga, An Ancient Swamp Witch Who Loves to Eat People." *Broadly*.
11. Herstik, Gabriela. (January 30, 2016) "The Cunning Female Demons and Ghosts of Ancient Japan." *Broadly*.
12. Enlow, Courtney. (October 18, 2017) "Chosen One of the Day: The Woman in Room 237 in The Shining." *SYFY Wire*.

CHAPTER EIGHTEEN: FRIDAY THE 13TH

1. Tate, James M. (June 1, 2015) "Ballet of the Machete: Friday the 13th Star Adrienne King." *Cult Film Freak*.
2. Divers, Anthony. (March 2019) "Pamela Voorhees: Nothing is Stronger Than a Mother's Love." *25 Years Later*.
3. (February 26, 2019) "Preventing Intimate Partner Violence Fact Sheet." *Centers for Disease Control and Prevention*.

4. Jackson, Matthew. (July 13, 2018) "17 Surprising Facts About Friday the 13th." *Mental Floss.*

5. (April 2019) "Hydrocephalus Fact Sheet." *National Institute of Neurological Disorders and Stroke.*

6. Lino, Mark. (August 2014) "Expenditures on Children by Families, 2013." *United States Department of Agriculture.*

7. Robg, Mike C. (May 2004) "Betsy Palmer Interview." *Icons of Fright.*

8. Pawlowski, A. (December 5, 2017) "Why Older Women Will Rule the World: The Future is Female, MIT Expert Says." *NBC News.*

CHAPTER NINETEEN: STRANGER THINGS

1. Xiong, Jesse Hong. (2010) *The Outline of Parapsychology.* University Press of America.

2. Barlow, Susanna. (September 18, 2017) "Understanding the Healer Archetype." *Thresholds.*

3. Howell, Elizabeth. (May 10, 2018) "Parallel Universes: Theories and Evidence." *Space.com.*

4. Bailey, Michael David. (2002) "The Feminization of Magic and the Emerging Idea of the Female Witch in the Late Middle Ages." *Essays in Medieval Studies.*

5. Fox, Margaret. (1954) "Searchlight on Psychical Research." *Rider and Company.*

6. Ulanov, Ann and Barry. (2015) *The Witch and the Clown: Two Archetypes of Human Sexuality.* Chiron Publications.

7. Kushner, Dale M. (May 31, 2016) "Mothers, Witches, and the Power of Archetypes." *Psychology Today.*

CHAPTER TWENTY: MISERY

1. Ebert, Roger. (November 30, 1990) "Misery." *Chicago Sun Times.*

2. (2019) "Bone Healing." Foothealthfacts.org.

3. Serena, Katie. (February 26, 2018) "Inside the Twisted Mind and Murders of Jolly Jane Toppan." *All That's Interesting.*

4. (2019) "Jane Toppan." *crimemuseum.org.*

5. Yardley, Elizabeth, and Wilson, David. (January 1, 2016) "In Search of the 'Angels of Death': Conceptualising the Contemporary Nurse Healthcare Serial Killer." *Journal of Investigative Psychology & Offender Profiling.*

6. Linic, Claire. (October 13, 2017) "The Truth Behind the Babysitter and the Man Upstairs Urban Legend." *Medium.*

7. Hur, Johnson. (November 26, 2016) "History of Babysitting." *bebusinessed. com.*

8. Forman-Brunell, Miriam. (2011) *Babysitter: An American History.* NYU Press.

9. Berry, Craig. (March 30, 2018) "The Unsolved Murder of Janett Christman." *True Crime Articles.*

10. Ericksson, Kimmo, and Coultas, Julie C. (March 1, 2014) "Corpses, Maggots, Poodles and Rats: Emotional Selection Operating in Three Phases of Cultural Transmission of Urban Legends." *Journal of Cognition & Culture.*

11. Bever, Lindsey. (26 May 2017) "'Angel of Death' Nurse Charged With Killing Another Baby, Suspected in up to 60 Other Deaths". *Washington Post.*

CHAPTER TWENTY-ONE: NAILS

1. Wixson, Heather. (November 17, 2017) "Interview: CoWriter/Director Dennis Bartok on Tapping into the Horrors of Nails with Shauna Macdonald." *Daily Dead.*

2. Couch, Stacey L.L. (December 29, 2016) "Archetypes: Healer and Wounded Healer." *Wild Gratitude.*

3. (2006) "22 'Heroic Deaths' by Black Characters in Horror Movies." *black-horrormovies.com.*

4. Resnick, Brian. (January 10, 2012) "What America Looked Like: Polio Children Paralyzed in Iron Lungs." *The Atlantic.*

CHAPTER TWENTY-TWO: THE HAUNTING OF HILL HOUSE

1. Shirleyjackson.org. (2009)

2. Tobey, Tas. (Oct 11, 2018) "Before Watching 'The Haunting of Hill House,' Read These 13 Haunted Books." *New York Times.*

3. Geggel, Laura. (October 26, 2015) "Scared to Death: Can You Really Die of Fright?" *Live Science.*

4. Stafford, Jeff. (June 4, 2019) "The Haunting." *Turner Classic Movies.*

5. Rubenstein, Roberta. (Autumn, 1996) "House Mothers and Haunted Daughters: Shirley Jackson and Female Gothic." *Tulsa Studies in Women's Literature.*

6. Ibid.

7. Solo, Andre. (January 18, 2019) "13 Signs That You're an Empath." *Highly Sensitive Refuge.*

8. Craig, Stephanie F. (December 2012) "Ghosts of the Mind: Supernatural and Madness in Victorian Gothic Literature." *University of Southern Mississippi Aquila Digital Community.*

9. Sarkis, Stephanie A., PhD. (January 22, 2017) "11 Warning Signs of Gaslighting." *Psychology Today.*

10. Zettel, Sarah. (September 10, 2018) "Are We All Gas Lighters? How Crime Fiction Helps Us Understand the Part Communities Play in Continuing Abuse." *All Crime Reads.*

CHAPTER TWENTY-THREE: HEREDITARY

1. Johnson, Greg. (Fall 1989) "Gilman's Gothic Allegory: Rage and Redemption in 'The Yellow Wallpaper.'" *Studies in Short Fiction.*
2. Todd, Carolyn L. (January 26, 2018) "Stranger Things Star Gaten Matarazzo Shares What Life Is Like With His Rare Bone Disorder." *Self.*
3. Bush, Erin N. (2019) "The Nutshell Studies of Unexplained Death." DeathinDiorama.com.
4. Tasca, Cecilia, et al. (October 19, 2012) "Women And Hysteria in the History Of Mental Health." *Clinical Practice & Epidemiology in Mental Health.*
5. (April 15, 2019) "Female Hysteria during Victorian Era: Its Symptoms, Diagnosis & Treatment/Cures." *Victorian Era.org.*
6. Wallace, Wendy. (May 12, 2012) "Sent to the Asylum: The Victorian Women Locked Up because They Were Suffering from Stress, Postnatal Depression and Anxiety." *Daily Mail.*
7. Sokol, Tony. (October 21, 2018) "Hereditary: The Real Story of King Paimon." *Den of Geek.*

CHAPTER TWENTY-FOUR: THE OTHERS

1. Jolluck, Katherine R. (December 1, 2016) "Women in the Crosshairs: Violence Against Women during the Second World War." *Australian Journal of Politics & History.*
2. (June 26, 2018) "Xeroderma Pigmentosum." *Genetics Home Reference.*
3. Excell, Jon. (December 22, 2015) "The Lethal Effects of London Fog." *BBC.*
4. Pegg, Samantha. (December 2009) "Madness is a Woman: Constance Kent and Victorian Constructions of Female Insanity." *Liverpool Law Review.*

CHAPTER TWENTY-FIVE: GHOSTBUSTERS

1. Sokol, Joshua. (July 11, 2016) "The MIT Physicists Who Infused Ghostbusters With Real Science." *Wired.*
2. Proctor, William. (2017) "'Bitches Ain't Gonna Hunt No Ghosts' Totemic Nostalgia, Toxic Fandom and the Ghostbusters Platonic." *Palabra Clave.*
3. Adams, Sam. (July 14, 2016) "Why the Ghostbusters Backlash is a Sexist Control Issue." *Indiewire.*
4. Ibid.
5. Dvorak, Petula. (July 14, 2016) "Ghostbusters, the Bros Who Hate It and the Art of Modern Misogyny." *Washington Post.*
6. Donaldson, Kayleigh. (October 2, 2017) "Women Love Horror: Why Does This Still Surprise So Many Dudes?" *SyFy Wire.*
7. Latchman, David. (2016) "The Science of the Ghostbuster's Upgraded Proton Pack." *Science vs Hollywood.*

8. (September 22, 2016) "Geena Davis Inclusion Quotient The Reel Truth: Women Aren't Seen or Heard." *Southern California Public Radio.*
9. Ibid.

CHAPTER TWENTY-SIX: THE X-FILES

1. Gates, Philippa. (2004) "Manhunting: The Female Detective in the Serial Killer Film." *Post Script.*
2. Rojas, Alejandro. (August 2, 2017) "New Survey Shows Nearly Half of Americans Believe in Aliens." *Huffington Post.*
3. (2019) "The Scully Effect. I Want to Believe in STEM." *Geena Davis Institute on Gender Media.*
4. (April 29, 2019) "Alien Abduction: an Unlikely Solution to the Climate Crisis." *The Guardian.*
5. Nyren, Erin. (October 17, 2018) "Jason Blum Says He's Meeting with Women Directors After Claiming 'There Aren't a Lot.'" *Variety.*

CHAPTER TWENTY-SEVEN: GINGER SNAPS

1. Allan, Kerri. (2001) "Katharine Isabelle." *Sci-Fi Online.*
2. Lawrence, Elizabeth A. (1996) "Werewolves in Psyche and Cinema: Man-Beast Transformation and Paradox." *Journal of American Culture.*
3. Wendelin, D., Pope, D., and Mallory, S. (2003) "Hypertrichosis." *Journal of the American Academy of Dermatology.*
4. Cininas, Jazmina. (2009) "Beware the Full Moon: Female Werewolves and 'That Time of the Month.'" *Grotesque Femininities.*
5. Clark, Elizabeth M. (April 2008) "Hairy Thuggish Women: Female Werewolves, Gender, and the Hoped-For Monster." *Georgetown University.*
6. Ingless-Arkell, Esther. (December 12, 2012) "The Science of Human Tails." *Gizmodo.*

CHAPTER TWENTY-EIGHT: BUFFY THE VAMPIRE SLAYER

1. Ferguson, C. J. (2012) "Positive Female Role-Models Eliminate Negative Effects of Sexually Violent Media." *Journal of Communication.* 62(5), pp. 888–899.
2. Ewens, Hannah. (May 7, 2017) "The Buffy Episode That Changed the Way We Talk About Television." *Vice.*
3. (January 3, 2018) "All Time Historical Women's Powerlifting World Records in Pounds/Kilograms." *Powerliftingwatch.com.*
4. Grady, Constance. (March 10, 2017) "Buffy the Vampire Slayer's Feminism Is Still Subversive, 20 Years Later." *Vox.*
5. Fredrickson, Karen. (2017) "Sarah Michelle Gellar's Stunt Double Talks Break Ground and Buffy's Fighting Style." *Music in the Dark.*

6. Wellons, Nancy Imperial. (February 13, 2001) "Waitman a Hit Doing Stunts for Gellar on Buffy." *Chicago Tribune.*

7. Sharpiro, Lila. (July 3, 2018) "Marti Noxon on Sharp Objects, Joss Whedon, and Going Toe-To-Toe With Jean-Marc Vallée." *Vulture.*

CHAPTER TWENTY-NINE: A QUIET PLACE

1. Thomson, Rosemarie Garland. (1997) "Feminist Theory, the Body, and the Disabled Figure." *The Disability Studies Reader.*

2. Squires, John. (March 14, 2018) "John Krasinski on the Importance of Casting Deaf Actress Millicent Simmonds in A Quiet Place." *Bloody Disgusting.*

3. Oates, Chris. (June 9, 2003) "How Were the Speed of Sound and the Speed of Light Determined and Measured?" *Scientific American.*

4. Datta, Sreela. (April 24, 2018) "Animals That Use Echolocation." *Sciencing.*

5. Talkington, Marilee. (October 26, 2010) "An Actor Finds Truth and Power Negotiating Her Vision Loss." *Brains of Minerva.*

6. Woodburn, Danny. (July 2016) "The Ruderman White Paper on Employment of Actors With Disabilities on Television." *The Ruderman White Foundation.*

7. Peacock, Addison. (October 21, 2016) "Hush Star Kate Siegel on Ouija, Women in Horror, and Understanding Privilege." *The Mary Sue.*

8. Thurman, Trace. (April 7, 2016) "Hush Director Mike Flanagan and Actress Kate Siegel on Their New Thriller." *Bloody Disgusting.*

9. Elliot, Laura. (November 11, 2018) "What's So Scary About Disability?" *The Establishment.*

10. Smith, Dr. Stacy L. (September 2016) "Inequality in 800 Popular Films: Examining Portrayals of Gender, Race/Ethnicity, LGBT, and Disability from 2007–2015." *The Annenberg Foundation.*

CHAPTER THIRTY: KICK-ASS WOMEN OF THE PAST AND FUTURE

1. Bartyzel, Monika. (October 31, 2014) "Girls on Film: 4 Female Filmmakers Who Pioneered Horror Movies." *The Week.*

2. Lunden, Jeff. (November 7, 2018) "King Kong on Broadway Is the 2,400-Pound Gorilla in the Room." *NPR.*

3. Roth, Dany. (March 6, 2017) "Kong Retrospective: King Kong 1933." *SYFY Wire.*

4. Kit, Borys. (November 3, 2017) "How Guillermo del Toro's Black Lagoon Fantasy Inspired Shape of Water." *The Hollywood Reporter.*

5. Del Toro, Guillermo. (January 18, 2019) "Millicent Patrick—a huge talent that was railroaded by egos and envies that surrounded her. The Gillman is #1 monster suit in film history, followed by Giger's Xenomorph." *Twitter.*

6. (October 2, 2016) "Creature From the Black Lagoon Fun Facts." *The Grindhouse Cinema Database.*

7. Luers, Erik. (June 15, 2016) "Mary Lambert on Pet Sematary, Non-Linear Narratives, and Child Actors." *Filmmaker Magazine.*

8. Krischer, Hayley. (October 30, 2014) "A Battle to the Grave: An Interview with the Soska Sisters." *The Hairpin.*

9. Thomas, Lou. (April 6, 2017) "Raw Director Julia Ducournau: 'I'm Fed Up with the Way Women's Sexuality Is Portrayed on Screen.'" *British Film Institute.*

INDEX